DICTIONARY
OF
SURFACE ACTIVE AGENTS,
COSMETICS AND TOILETRIES

ENGLISH, FRENCH, GERMAN, SPANISH, ITALIAN, DUTCH, POLISH

by

Dr. G. CARRIÈRE

Unilever Research, Vlaardingen/Duiven,
Vlaardingen, The Netherlands

ELSEVIER SCIENTIFIC PUBLISHING COMPANY
AMSTERDAM — OXFORD — NEW YORK
1978

Published in co-edition with WNT, Warsaw

The Polish equivalents and definitions have been added
by P. KIKOLSKI
Institute of Industrial Chemistry
Department of Household Chemistry, Warsaw

Distribution of this book is handled by the following publishers :
    for the U. S. A. and Canada

Elsevier/North–Holland, Inc.
52 Vanderbilt Avenue
New York, N. Y. 10017
    for Poland

Wydawnictwa
Naukowo-Techniczne,
P. O. Box 350
00-950 Warszawa, Poland
    for all remaining areas

Elsevier Scientific Publishing Company
335 Jan van Galenstraat
P. O. Box 211, Amsterdam, The Netherlands

Library of Congress Cataloging in Publication Data
Carrière, Gerardus.
    Dictionary of surface active agents, cosmetics, and toiletries.

    Includes indexes.
    1.   Surface active agents—Dictionaries—Polyglot.
2.   Cosmetics—Dictionaries—Polyglot.   3.   Toilet
preparations—Dictionaries—Polyglot.   4.   Dictionaries,
Polyglot.   I. Kikolski, P.   II. Title.   III. Title:
Dictionary of surface active agents, cosmetics and
toiletries.
TP994.C37 1977   668'.1'03   77-8552
ISBN 0-444-99809-8

DICTIONARY OF
SURFACE ACTIVE AGENTS,
COSMETICS AND TOILETRIES

PREFACE

It is now nearly a quarter of a century since I first realised how many new terms and specific expressions the modern industry of detergents, cosmetics and toiletries had introduced.

Bearing in mind my own difficulties in choosing the appropriate phrases even in my own language, I conceived the idea of a classified glossary covering the principal modern languages, which would help me when writing letters to colleagues, and when listening to talks at international conferences. I sollicited the help of friends and experts and as a result a first glossary of terms used in the detergents industry was published in a journal in 1958.

Following this first version, which covered three languages only, I was encouraged to extend the idea and other versions covering more languages followed. Finally in 1960 the collection was published in book form as a nineteen-language glossary by the Elsevier Publishing Company, followed in 1966 by an edition covering the closely allied field of cosmetics and toiletries as well.

In the meantime the International Commission of Terminology (C.I.T.) of the Comité International des Dérivés Tensio-Actifs (C.I.D.) has studied and is still studying the terms and concepts used in the field of surface phenomena, and, among my many obligations, I am particularly indebted to my friends and colleagues of the C.I.T. for their ready permission to make use of their definitions. I particularly wish to thank:

Dr. W. Langmann, President (Germany);
Dr. L. Lascaray (Spain);
Dr. G. Pozzi (Italy) and
Mr. C. Kortland (Netherlands).

Furthermore I also acknowledge my indebtedness both to the International Organization for Standardization (ISO) and the International Union of Pure and Applied Chemistry (IUPAC) for permission to call attention to the major divergences of their definitions from the C.I.T. definitions.

My thanks are also due to my indefatigable secretary, Mrs. P. den Hollander, for preparing the whole of the manuscript.

Finally, I wish to thank my wife for her never-failing interest and encouragement. She has helped me to appreciate the truth of that touching phrase in the conclusion of Odysseus's first speech to Nausikaa:

ὁμοφρονέοντε νοήμασιν
(Odyssey VI, 183)

BASIC TABLE

A

1 abrading toothpowder
fr poudre dentifrice abrasive
de abschleifendes Zahnpulver
el polvo dentífrico abrasivo
it polvere dentifricia abrasiva
ne schurend tandpoeder
pl proszek do zębów o działaniu ściernym

2 abrasive
fr abrasif
de Schleifmittel, Putzkörper
el abrasivo
it abrasivo
ne schuurmiddel
pl materiał ścierny

3 acid cream
fr crème acide
de saure Creme
el crema ácida
it crema acida
ne zure crème
pl krem kwaśny, krem o pH równym pH skóry

4 acid layer of the skin, acid mantle of the skin
fr couverture acide de la peau, revêtement acide de la peau, manteau acide protecteur de la peau
de Säureschutzhülle, Säuremantel der Haut
el cobertura ácida de la piel, revestimiento de la piel, capa protectora ácida de la piel
it strato acido della pelle, rivestimento acido della pelle
ne zuurmantel van de huid
pl warstwa kwasowa skóry, kwaśna ochronna warstwa powierzchniowa skóry, kwasowy płaszcz powierzchniowy skóry

5 acid mantle
fr manteau acide
de Säuremantel
el capa ácida
it tonico (per la pelle)
ne zuurmantel
pl płaszcz kwasowy

6 active matter
In a composition, the whole of the specific constituents responsible for the activity specified.
Note: The specific constituents are determined by analytical criteria.

fr matière active
Dans une composition, ensemble des constituants spécifiques responsables d'une action déterminée.
Nota: Les constituants spécifiques sont définis par les critères analytiques.
de Aktivsubstanz
In einer Mischung grenzflächenaktiver Verbindungen die Gesamtheit der wesentlichen Bestandteile, welche eine gewünschte Wirkung hervorrufen.
Anmerkung: Die praktische Definition der Aktivsubstanz erfolgt nach analytischen Kriterien.
el materia activa
En una composición: conjunto de componentes específicos, responsables de una acción determinada.
Observación: Los componentes específicos quedan definidos por criterios analíticos.
it sostanza attiva
Il o i costituenti specifici di una miscela aventi un'azione determinata.
Nota: I costituenti specifici sono definiti con criteri analitici.
ne aktieve stof
In een compositie, het geheel van die produkten van de samenstelling die bepalend zijn voor een bepaalde eigenschap.
Opmerking: De specifieke bestanddelen worden bepaald door analytische kriteria.
pl substancja aktywna
W gotowej kompozycji zespół składników odpowiedzialnych za określone działanie powierzchniowe.
Uwaga: Substancję aktywną określa się według kryteriów analitycznych.

7 active oxygen, oxygen release
fr libération de l'oxygène
de Freiwerden von Sauerstoff, Sauerstoffabspaltung
el liberación del oxígeno
it liberazione di ossigeno
ne vrijkomen van de zuurstof
pl uwalnianie aktywnego tlenu

8 adsorption layer of surface active agents
In the case of surface active substances in solution: a layer stretching more or less across an interface and the thickness of which is determined by the fact that at any random place of that layer, the concentration of the adsorbed product is greater than that in each of the contiguous phases.

fr couche d'adsorption d'agents de surface
Dans le cas de solutions d'agents de surface: couche s'étendant plus ou moins de part et d'autre d'une interface et dont l'épaisseur est déterminée par le fait qu'en un lieu quelconque de cette couche, la concentration d'un produit adsorbé est supérieure à celle existant dans chacune des phases avoisinantes.

de Grenzflächenschicht (nur für Tensidlösungen)
Schicht, die sich ganz oder teilweise auf einer Grenzfläche ausbreitet und deren Dicke dadurch bestimmt ist, dass in ihr die Konzentration der adsorbierten Substanz grösser ist als in jeder der beiden benachbarten Phasen.

el capa de adsorción
En el caso de disoluciones de agentes de superficie: capa que se extiende más o menos a ambos lados de una interfacie y cuyo espesor está determinado porque, en un lugar cualquiera de esta capa, la concentración del producto adsorbido es mayor que la que existe en cada una de las fases contiguas.

it strato di assorbimento
Nel caso dei tensioattivi in soluzione, strato estendentesi più o meno da una parte e dall'altra di una interfaccia e il cui spessore è individuato dalla concentrazione del prodotto assorbito, superiore in ciascun punto dello strato a quella esistente nelle fasi contigue.

ne adsorptielaag
In geval van oppervlakaktieve stoffen in oplossing: een laag die zich min of meer over een grensvlak uitstrekt en waarvan de dikte bepaald wordt door het feit dat op elke willekeurige plaats van de laag de concentratie van het geadsorbeerde produkt groter is dan in die van de kontinue fasen.

pl warstwa adsorpcyjna związków powierzchniowo czynnych
W przypadku związków powierzchniowo czynnych występujących w roztworze jest to warstewka występująca na granicy faz, której grubość jest określona tak, że stężenie w każdym punkcie warstewki jest większe niż w każdej z dwu sąsiadujących faz.

9 advancing wetting angle
The introduction of a solid surface at slow and constant speed into a liquid phase gives rise to the formation of a contact angle the dimension of which may depend upon the nature of the surface and the speed of entry. This angle is called the advancing wetting angle.

Notes: 1. As generally measured, the advancing wetting angle is that which corresponds to penetration perpendicular to the surface of the liquid phase.
2. In the case of a solid surface, the advancing and receding wetting angles relating to a liquid phase are different. The difference arises from measurement of the angles produced at the speed of introduction and withdrawal laid down in the test method.

fr angle de mouillage rentrant
La pénétration lente et à vitesse constante d'une surface solide vers l'intérieur d'une phase liquide donne lieu à la formation d'un angle de contact dont la grandeur peut dépendre de la nature de la surface et de la vitesse de pénétration. Cet angle est appelé angle de mouillage rentrant.

Nota: 1. Un angle de mouillage rentrant, généralement mesuré, est celui qui correspond à une pénétration effectuée perpendiculairement à la surface de la phase liquide.
2. Dans le cas d'une surface solide, les angles de mouillage rentrant et sortant vis-à-vis d'une phase liquide sont différents. La différence résulte de mesures d'angles effectuées dans des domaines de vitesse de pénétration et d'extraction, précisées par les méthodes d'essais.

de fortschreitender Randwinkel
Das langsame und mit konstanter Geschwindigkeit erfolgende Eintauchen einer Festkörperoberfläche in das Innere einer flüssigen Phase bedingt die Bildung eines Kontaktwinkels, dessen Grösse von der Natur der Oberfläche und der Eintauchgeschwindigkeit abhängen kann. Dieser Winkel wird als fortschreitender Randwinkel bezeichnet.

Anmerkungen: 1. Ein im allgemeinen gemessener fortschreitender Randwinkel ist derjenige, der beim Eintauchen senkrecht zur Oberfläche der flüssigen Phase bestimmt werden kann.
2. Im Falle einer festen Oberfläche sind die fortschreitenden und rückläufigen Randwinkel verschieden. Die Differenz resultiert aus Messungen der Winkel in einem Geschwindigkeitsgebiet des Eintauchens und Herausziehens, die durch Testmethoden präzisiert werden.

el ángulo de mojado entrante
La penetración lenta y a velocidad constante de una superficie sólida hacia el interior de una fase líquida da lugar a la formación de un ángulo de contacto, cuyo valor puede depender de la naturaleza de la superficie y de la velocidad de penetración. Este ángulo se llama ángulo de mojado entrante.

Observaciones: 1. El ángulo de mojado entrante, el más comúnmente medido, corresponde a una penetración efectuada perpendicularmente a la superficie de la fase líquida.
2. En el caso de una superficie sólida, los ángulos de mojado entrante y saliente, con respecto a una fase líquida, son diferentes. La diferencia resulta de medidas efectuadas a las velocidades de penetración o de extracción prescritas en los métodos de ensayo.

it angolo di bagnatura progressiva
L'introduzione di una superficie solida in una fase liquida, a velocità lenta e costante, dà luogo alla formazione di un angolo di contatto che può dipendere dalla natura della superficie e dalla velocità di introduzione. Tale angolo è chiamato angolo di bagnatura progressiva.

Note: 1. L'angolo di bagnatura progressiva, come viene generalmente misurato, è quello che corrisponde ad una introduzione effettuata perpendicolarmente alla superficie della fase liquida.
2. Per una superficie solida l'angolo di bagnatura progressiva e l'angolo di bagnatura regressiva, relativi ad una fase liquida, sono differenti.
La differenza risulta dalle misure degli angoli effettuate alle velocità di introduzione e di estrazione precisate nei metodi di prova.

ne progressierandhoek
Wanneer een vast oppervlak met een geringe en konstante snelheid in een vloeibare fase wordt gebracht, ontstaat een randhoek afhankelijk van de aard van het oppervlak en de ingangssnelheid. Deze hoek wordt de progressierandhoek genoemd.

Opmerkingen: 1. Gewoonlijk wordt de progressierandhoek gemeten bij vertikale penetratie van het oppervlak van de vloeistoffen.
2. In het geval van een vast oppervlak is er verschil tussen progressie- en regressierandhoek ten opzichte van een vloeibare fase. Het verschil ontstaat door meting van de hoeken die ontstaan bij de snelheid van inbrengen en uitnemen welke is vastgelegd in de beproevingsmethode.

pl kąt zwilżania wstępujący
Wprowadzanie powolne i ze stałą prędkością powierzchni ciała stałego do fazy ciekłej powoduje powstanie kąta zwilżania, którego wartość może zależeć od charakteru powierzchni i szybkości zanurzania. Kąt ten nazywany jest wstępującym kątem zwilżania.

Uwagi: 1. Normalnie mierzony wstępujący kąt zwilżania występuje przy zanurzaniu prostopadle do powierzchni fazy ciekłej.
2. Dla danej powierzchni ciała stałego występuje różnica pomiędzy wstępującym i zstępującym kątem zwilżania. Różnica ta wynika z pomiarów kątów powstałych przy zanurzaniu i wyjmowaniu z szybkością podaną w metodzie badań.

10 advancing wetting tension
Wetting tension corresponding to the formation of an advancing wetting angle.

fr tension de mouillage rentrante
Tension de mouillage correspondant à la formation d'un angle de mouillage rentrant.
de fortschreitende Benetzungsspannung
Ist die einem fortschreitenden Randwinkel entsprechende Benetzungsspannung.
el tensión de mojado entrante
Tensión de mojado correspondiente a la formación de un ángulo de mojado entrante.
it tensione di bagnatura progressiva
Tensione di bagnatura corrispondente alla formazione di un angolo di bagnatura progressiva.
ne progressiebevochtigingsspanning
Bevochtigingsspanning die overeenkomt met de vorming van een progressiebevochtigingshoek.
pl napięcie zwilżające wstępujące
Napięcie zwilżające odpowiadające powstawaniu wstępującego kąta zwilżania.

11 aerosol foam waving compound
fr produit aérosol moussant pour ondulation
de Aerosol-Schaumwellprodukt
el compuesto aerosol espumante para la ondulación
it aerosol a schiuma per ondulazione
ne schuimend haargolfmiddel in aërosol
pl aerozolowy pieniący środek do ondulacji

12 aerosol toothpaste
fr pâte dentifrice aérosol
de Zahnpaste in Druckpackung
el pasta dentífrica aerosol
it pasta dentifricia aerosol
ne aërosol-tandpasta
pl aerozolowa pasta do zębów

13 affinity to fibres
fr affinité pour les fibres, pouvoir de monter sur la fibre
de Aufziehvermögen auf die Faser, Affinität zur Faser
el afinidad para las fibras
it affinità per la fibra, potere di montare sulla fibra
ne affiniteit tot de vezels
pl powinowactwo do włókien

14 after-shave lotion (preparation)
fr lotion après rasage, préparation après rasage

de Rasiermittel zur Nachbehandlung
le loción para después del afeitado, preparado para después del afeitado
it lozione dopo rasatura
ne gezichtswater voor na het scheren
pl płyn po goleniu

15 after-taste
fr après-goût, arrière-goût
de Nachgeschmack
el último sabor
it sapore dopo, sapore prima
ne nasmaak
pl posmak

16 after-treating agent for prints
Product covered by the definition
corresponding to the Dye-fixing agent.
fr agent pour le traitement subséquent des
impressions
Produit auquel s'applique la définition
correspondant à Agent pour le traitement
subséquent des teintures.
de Nachbehandlungsmittel für Drucke
Produkt, für das die Definition zu
"Nachbehandlungsmittel für Färbungen"
zutrifft.
el agente para el tratamiento posterior de
estampados
Producto el cual se aplica la definición
correspondiente a Agente para el tratamiento posterior de tinturas.
it —
ne nabehandelingsmiddel voor bedrukte
stoffen
Een produkt waarop de definitie van
het verffixeermiddel van toepassing is.
pl środek utrwalający dla wydruków
Produkt objęty definicją środka po-
mocniczego do utrwalania wybarwień.

17 all-purpose cream
fr crème tous usages
de Allzweckcreme
el crema para todo uso
it crema per tutti gli usi
ne universele crème
pl krem uniwersalny

18 all-purpose washing agent
fr produit universel de nettoyage
de Universalreinigungsmittel
el producto universal de limpiar, detergente para todo
it detergente polivalente
ne universeel reinigingsmiddel
pl uniwersalny środek do prania

19 alpha phase
The crystalline form of slightly hydrated
soap which reverts to the beta phase
form through complete dehydratation.
This phase does not exist in the conditions normally encountered in commer-
cial soaps.
fr phase alpha
Forme cristalline du savon peu hydraté,
qui reverse à la forme de phase bêta
par déshydratation complète. Cette phase
n'existe pas dans les conditions normalement rencontrées dans les savons com-
merciaux.
de Alpha-Phase
Kristalline Form niedrig hydratisierter
Seife, die durch vollständige Dehydra-
tisierung in die Beta-Phase übergeht.
Diese Phase ist in Handelsseifen normalerweise nicht enthalten.
el fase alfa
Forma cristalina del jabón poco hidra-
tado, que vuelve a la forma de fase beta
por deshidratación completa. Normalmente, esta fase no existe en los jabones
comerciales.
it fase alfa
Forma cristalina del sapone poco
idratato, che può passare alla fase beta
attraverso la completa deidratazione.
Questa fase non esiste normalmente nei
saponi commerciali.
ne alpha-fase
Kristallijnen vorm van weinig gehy-
drateerde zeep, die door volledige dehy-
dratatie in de beta-fase overgaat.
pl faza alfa
Krystaliczna postać lekko uwodnionego
mydła przechodząca przy pełnym od-
wodnieniu w fazę beta. Faza ta normalnie nie występuje w mydłach dostęp-
nych w handlu.

20 amide formation
Chemical reaction giving rise to the
formation of amides by the action of
ammonia or primary or secondary
amines on acids, their halides or their
esters.
fr amidification
Réaction chimique permettant d'obtenir
des amides par l'action soit d'ammoniac, soit d'amines primaires ou secondaires, sur des acides, leurs halogénures
ou leurs esters.

de Amidierung
Chemische Reaktion, bei welcher Säure-
amide entstehen durch Einwirkung
von Ammoniak, primären oder sekun-
dären Aminen auf Säuren, ihre Halo-
genide oder Ester.

el amidificación
Reacción química, que permite obtener
amidas por la acción del amoniaco,
o de aminas primarias o secundarias,
sobre ácidos, sus halogenuros o sus
ésteres.

it amidificazione
Reazione chimica che porta alla formazione di amidi per azione d'ammoniaca o ammine primarie o secon-
darie su'acidi, su loro alogenuri o estari.

ne amidering
Chemische reactie waarbij amiden ontstaan door inwerking van ammoniak of
primaire resp. secundaire aminen op
zuren, hun halogeniden of hun esters.

pl amidowanie
Reakcja chemiczna, w wyniku której
poprzez działanie amoniaku lub pierw-
szo- względnie drugorzędowej aminy
na kwas, jego halogenek lub ester
powstaje amid kwasu.

21 amino acid
fr acide aminé
de Aminosäure
el amino-ácido
it aminoacido
ne aminozuur
pl aminokwas

22 amino dye
fr colorant aminé
de Aminofarbstoff
el colorante aminado
it colorante amminico
ne aminokleurstof
pl barwnik aminowy

23 ammoniated toothpaste
fr pâte dentifrice à l'ammoniaque
de Ammoniumzahnpaste
el pasta dentífrica al amoníaco
it pasta dentifricia all'ammoniaca
ne tandpasta met ammoniumverbinding
pl amoniakalna pasta do zębów

24 ammonium thioglycolate
fr thioglycolate d'ammonium, thioglycolate d'ammoniaque

de Ammoniumthioglykolat
el tioglicolato amónico
it tioglicolato di ammonio
ne ammoniumthioglycolaat
pl tioglikolan amonu

25 amount of foam
fr quantité de mousse
de Schaummenge
el cantidad de espuma
it quantità di schiuma
ne hoeveelheid schuim
pl ilość piany

26 amphiphilic product
Product comprising in its molecule, at
the same time, one or more hydrophilic
groups and one or more hydrophobic
groups.
Note: Surface active agents are amphiphilic products.

fr produit amphiphile
Produit renfermant dans sa molécule
à la fois un ou des groupements hydrophiles et un ou des groupements lipophiles.
Nota: Les agents de surface sont des produits
amphiphiles.

de amphiphiles Produkt
Produkt, das in seinem Molekül gleich-
zeitig hydrophile und hydrophobe
Gruppen enthält.
Anmerkung: Alle grenzflächenaktive Stoffe sind
amphiphile Produkte.

el producto anfifilo
Compuesto que contiene en su molécula,
a la vez, uno o varios grupos hidrófilos
y uno o varios grupos lipófilos.
Observación: Los agentes de superficie son productos anfifilos.

it composto anfifilo
Composto la cui molecola contiene uno
o più gruppi idrofili insieme a uno
o più gruppi lipofili.
Nota: I tensioattivi sono composti anfifili.

ne amfifiele verbinding
Verbinding waarbij het molecuul tegelijkertijd een of meer hydrofiele groepen
en een of meer lipofiele groepen bevat.
Opmerking: Oppervlakaktieve stoffen zijn amfifiele verbindingen.

pl związek amfifilny
Związek zawierający w swej cząsteczce
jednocześnie jedną lub więcej grup
hydrofilowych oraz jedną lub więcej
grup hydrofobowych.
Uwaga: środki powierzchniowo czynne są produktami amfifilnymi.

27 **ampholytic surface active agent**
A surface active agent having two or more functional groups which, depending on the conditions of the medium, can be ionized in an aqueous solution and give to the compound the characteristics of an anionic or a cationic surface active agent. This ionic behaviour is similar to that of amphoteric compounds in the broadest sense.

fr agent de surface ampholyte
Agent de surface possédant deux ou plusieurs groupements fonctionnels qui peuvent, selon les conditions du milieu, s'ioniser en solution aqueuse, en conférant au composé le caractère d'agent de surface anionique ou cationique. Ce comportement ionique est analogue à celui des composés amphotères au sens le plus général.

de ampholytische grenzflächenaktive Verbindung
Grenzflächenaktive Verbindung, die zwei oder mehrere funktionelle Gruppen besitzt, die in wässriger Lösung ionisieren können und dabei — je nach Bedingungen des Mediums — dem Produkt anionischen oder kationischen Charakter verleihen. Dieses ionogene Verhalten ist im weitesten Sinne analog dem der amphoteren Verbindungen.

el agente de superficie anfólito
Agente de superficie, que posee dos o más grupos funcionales, que pueden ionizarse en disolución acuosa, confiriendo al producto, según las condiciones del medio, el carácter de agente de superficie aniónico o catiónico. Este comportamiento iónico es análogo al de los compuestos anfóteros en su sentido más amplio.

it tensioattivo anfolito
Tensioattivo con due o più gruppi funzionali che possono ionizzarsi in soluzione acquosa e dare al composto caratteristiche di tensioattivo anionico o cationico. Questo comportamento ionico è analogo a quello dei composti anfoteri nel senso più generale.

ne amfotere stof
Een oppervlakaktieve stof die één of meer funktionele groepen bevat die, afhankelijk van de toestand van het medium, in een waterige oplossing geïoniseerd kunnen worden, waardoor de verbinding de eigenschappen van een anionaktieve of kationaktieve stof aanneemt. Dit ionisch gedrag komt overeen met dat van amfionaktieve verbindingen in de ruimste zin.

pl amfolityczny związek powierzchniowo czynny
Związek powierzchniowo czynny posiadający dwie grupy funkcyjne, które zależnie od warunków środowiska mogą ulegać jonizacji w roztworze wodnym dając związek powierzchniowo czynny o własnościach anionowych lub kationowych. Takie własności jonowe są analogiczne — w najszerszym znaczeniu — do własności związków amfoterycznych.

28 **ampholytics**
fr composés ampholytes
de Ampholyte
el compuestos anfólitos
it composti anfoliti
ne amfifiele stoffen
pl amfolity

29 **amphoteric surfactants**
fr surfactifs amphotères
de amphotere oberflächenaktive Verbindungen
el compuestos anfotéricos
it composti tensioattivi anfoteri
ne amfionaktieve stoffen
pl amfoteryczne związki powierzchniowo czynne

30 **analgesic**
fr analgésique
de Analgetikum
el analgésico
it analgesico
ne pijnstillend middel
pl środek przeciwbólowy

31 **ancillary (for surface active agents)**
A complementary component of a detergent which imparts added properties unrelated to the washing action as such. Examples: optical bleaches, corrosion inhibitors, antielectrostatic agents, colouring matters, perfumes, bactericides, etc.
Note: Ancillaries are usually present in small quantities.

fr additif (pour agents de surface)
Composant complémentaire d'un détergent qui introduit des propriétés

étrangères à l'action spécifique du lavage. Exemples: agents de blanchiment optique, inhibiteurs de corrosion, agents antiélectrostatiques, colorants, parfums, bactéricides, etc.

Nota: Les additifs interviennent généralement en faible quantité.

de Zusatzstoff (für Tenside)
Zusätzlicher Bestandteil eines Waschmittels, welcher weitere Eigenschaften einführt, die mit der Waschwirkung nichts zu tun haben. Beispiele: Optische Aufheller, Korrosionsinhibitoren, Antielektrostatika, Farbstoffe, Riechstoffe, Bakterizide, etc.

Anmerkung: Zusatzstoffe sind im allgemeinen in kleinen Mengen vorhanden.

el aditivo (para agentes de superficie)
Componente complementario de un detergente, que aporta propiedades ajenas a la acción específica del lavado. Ejemplos: blanqueadores ópticos, inhibidores de corrosión, agentes antielectrostáticos, colorantes, perfumes, bactericidas, etc.

Observación: Los aditivos intervienen generalmente en débil proporción.

it additivo
Costituente complementare di un detergente che gli conferisce altre proprietà oltre a quelle dell'azione specifica del lavaggio.

ne hulpstof (voor oppervlakaktieve stoffen)
Een complementair bestanddeel van een wasmiddel dat hieraan extra eigenschappen verleent die niet op de waswerking zelf betrekking hebben.
Voorbeelden: optische bleekmiddelen, anticorrosiemiddelen, antistatische middelen, kleurstoffen, parfums, bactericiden. Hulpstoffen zijn gewoonlijk in kleine hoeveelheden aanwezig.

pl dodatek pomocniczy (do związków powierzchniowo czynnych)
Dodatkowy składnik środków piorących nadający kompozycji, do której jest dodany, własności niezwiązane z działaniem piorącym. Przykłady: rozjaśniacze optyczne, inhibitory korozji, środki antyelektrostatyczne, barwniki, środki zapachowe, środki bakteriobójcze itd.

Uwaga: Dodatki pomocnicze obecne są na ogół w małych ilościach.

32 anionic surface active agent
A surface active agent which has one or more functional groups and ionizes in aqueous solution to produce negatively charged organic ions responsible for the surface activity.

fr agent de surface anionique
Agent de surface possédant un ou plusieurs groupements fonctionnels s'ionisant en solution aqueuse, pour fournir des ions organiques chargés négativement et responsables de l'activité de surface.

de anionische grenzflächenaktive Verbindung
Grenzflächenaktive Verbindung, die eine oder mehrere funktionelle Gruppen besitzt, die in wässriger Lösung ionisieren unter Bildung negativ geladener organischer Ionen, die für die Grenzflächenaktivität verantwortlich sind.

el agente de superficie aniónico
Agente de superficie que posee uno o varios grupos funcionales, que se ionizan en disolución acuosa, dando iones orgánicos cargados negativamente, responsables de la actividad de superficie.

it tensioattivo anionico
Tensioattivo con uno o più gruppi funzionali, che si ionizzano in soluzione acquosa formando ioni organici carichi negativamente e responsabili della tensioattività.

ne anionaktieve stof
Een oppervlakaktieve stof die één of meer funktionele groepen bevat en die in een waterige oplossing een ioniserende werking heeft en zodoende negatief geladen organische ionen produceert, die de oppervlakaktiviteit teweegbrengen.

pl anionowy związek powierzchniowo czynny
Związek powierzchniowo czynny zawierający jedną lub więcej grup funkcyjnych, ulegający jonizacji w roztworze wodnym z utworzeniem ujemnie naładowanych jonów odpowiedzialnych za aktywność powierzchniową.

33 anionics
fr composés anioniques
de anionaktive Verbindungen
el compuestos aniónicos
it composti anionici
ne anionaktieve stoffen
pl anionowe związki powierzchniowo czynne

34 antidandruff agent
fr produit antipelliculaire

de Antischuppenpräparat
el agente anti-caspa, agente contra la caspa
it prodotto antiforfora
ne middel tegen roos
pl preparat przeciwłupieżowy

35 antidandruff shampoo
fr shampooing antipelliculaire
de Antischuppenshampoo
el champu anti-caspa
it shampoo antiforfora
ne anti-roos shampoo
pl szampon przeciwłupieżowy

36 antielectrostatic agent[1]
Product which, when applied to a textile article during or after processing, makes it possible to eliminate the disadvantages due to phenomena of static electricity.
Note: These products are hydrophile or surface active agents, for example alcane sulphonates, alkylarylsulphonates, alkylphosphates, alkylamines and their derivatives and also the ethoxylation products of fatty acids, fatty alcohols, fatty amines, alkylphenols and onium salts.

fr agent antiélectrostatique[2]
Produit qui, appliqué à un article textile en cours d'élaboration ou terminé, permet d'éviter les inconvénients dus aux phénomènes d'électrisation.
Nota: Il s'agit de substances hydrophiles ou tensio-actives comme, par exemple: alkylsulfonates, alkylarylsulfonates, alkylphosphates, alkylamines et leurs dérivés, ainsi que des produits d'éthoxylation d'acides gras, d'alcools gras, d'amines grasses, d'amides gras, d'alkylphénols et des sels de dérivés onium.

de Mittel für die antielektrostatische Ausrüstung (Antielektrostatika[3])
Produkt, das auf einen textilen Rohstoff oder ein textiles Fertigerzeugnis aufgebracht wird, um Schwierigkeiten bei der Verarbeitung oder beim Tragen, die auf eine elektrostatische Aufladung zurückzuführen sind, auszuschalten.
Anmerkung: Es handelt sich um hydrophile oder grenzflächenaktive Stoffe, wie z. B. Alkansulfonate, Alkylarylsulfonate, Alkylphosphate, Alkylamine und ihre Derivate, Oxalkylierungsprodukte von Fettsäuren, Fettalkoholen, Fettaminen, Fettsäureamiden und Alkylphenolen sowie Oniumsalze.

el agente antielectrostático
Producto que se adiciona a una materia textil, en curso de elaboración o terminada, para evitar las dificultades debidas a la carga electrostática. Con frecuencia impropiamente llamado agente antiestático.
Observación: Se trata, en general, de agentes de superficie, como alquilsulfonatos, alquilfosfatos, alquilaminas y sus derivados, así como productos de oxialquilación de ácidos grasos, de alcoholes grasos, de aminas y amidas grasas, de alquilfenoles y de sales de amonio cuaternario.

it —
ne antistatikum
Avivage of bestanddeel van een avivage welke toegepast op vezel, garen of doek het vormen van electrostatische lading voorkomt of de gevormde lading doet afvloeien.

pl antystatyk
Produkt, który naniesiony na surowiec lub wyrób włókienniczy podczas procesu produkcyjnego lub już po wyprodukowaniu umożliwia wyeliminowanie niepożądanych zjawisk związanych z występowaniem ładunków elektryczności statycznej.
Uwaga: Produktami takimi są związki hydrofilowe lub związki powierzchniowo czynne takie jak alkanosulfoniany, alkiloarylosulfoniany, alkilofosforany, alkiloaminy i ich pochodne, produkty oksyetylenowania kwasów, alkoholi i amin tłuszczowych, produkty oksyetylenowania alkilofenoli oraz sole oniowe.

[1] Frequently incorrectly called "antistatic agent"†.
[2] Souvent, improprement appelé "agent antistatique"†.
[3] Oft, aber unrichtig, "Antistatika" genannt.
† See appendix
Voir appendice

37 anti-enzyme toothpaste
fr pâte dentifrice anti-enzyme
de anti-enzymatische Zahnpaste
el pasta dentífrica anti-enzimas
it pasta dentifricia anti-enzimatica
ne anti-enzyme tandpasta
pl antyenzymatyczna pasta do zębów

38 antifoaming agent, antifoamer
A substance which prevents the formation of a foam or considerably reduces foam persistence.
fr agent antimoussant, antimousse
Produit empêchant la formation d'une mousse ou en diminuant considérablement la stabilité.
de Entschäumer
Produkt, das die Bildung von Schaum verhindert oder seine Stabilität beträchtlich vermindert.
el producto antiespumante, antiespumante
Producto que impide la formación de espuma o disminuye mucho su estabilidad.
it antischiuma
Sostanza che impedisce la formazione di schiuma o ne diminuisce considerevolmente la stabilità.

ne antischuimmiddel
Een stof die het vormen van schuim verhindert of de stabiliteit van schuim aanzienlijk vermindert.

pl środek przeciwpieniący
Produkt zapobiegający tworzeniu się piany lub znacznie zmniejszający jej trwałość.

39 antifoaming agent for the textile industry
Product which prevents the formation of foam or which considerably reduces its stability. In the textile industry, it is used particularly in sizing, finishing and dye baths, in printing pastes, etc.
Note: These products include, among other things, certain surface active agents and preparations comprising them, for example those based on oils, phosphoric acid esters and alcohols of high molecular weight and their oxalkylated products.

fr agent antimousse pour l'industrie textile
Produit qui combat la formation d'une mousse ou qui diminue considérablement sa stabilité. Dans l'industrie textile, il est utilisé notamment dans les bains d'encollage, d'apprêts et de teinture, dans les pâtes d'impression, etc.
Nota: Il s'agit entre autres de certains agents de surface et de préparations en comportant, par exemple, à base d'huiles, d'esters de l'acide phosphorique, d'alcools à masse moléculaire élevée et de ses produits d'oxyalkylation.

de Schaumdämpfungsmittel für die Textilindustrie
Produkt, das die Bildung eines Schaumes verhindert oder seine Stabilität beträchtlich mindert. In der Textilindustrie wird es vorzugsweise z.B. in Schlicht- und Appreturflotten sowie in Färbebädern, Druckpasten usw. angewendet.
Anmerkung: Es handelt sich u.a. um bestimmte grenzflächenaktive Stoffe und Zubereitungen hieraus z.B. auf der Basis von Ölen, Phosphorsäureestern, höhermolekularen Alkoholen und deren Oxalkylierungsprodukten.

el agente antiespumante para la industria textil
Producto que atenúa la formación de espuma o que disminuye considerablemente su estabilidad. Se utiliza en la industria textil preferentemente en los baños de encolado, de apresto y de tintura, así como en las pastas de estampación, etc.
Observación: Se trata, entre otros, de determinados agentes de superficie y de preparaciones que los contienen, por ejemplo a base de aceites, de ésteres del ácido fosfórico y de alcoholes de elevado peso molecular.

it —

ne antischuimmiddel voor de textielindustrie
Een produkt dat schuimvorming voorkomt of de stabiliteit van het schuim aanmerkelijk vermindert. Het wordt in de textielindustrie vooral gebruikt in appreteer-, finishing- en verfbaden, in drukverven enz.

pl środek przeciwpieniący dla przemysłu włókienniczego
Produkt zapobiegający tworzeniu się piany lub znacznie zmniejszający jej trwałość. W przemyśle włókienniczym stosowany jest on głównie w kąpielach do klejenia, wykańczania i farbowania oraz w pastach drukarskich itd.
Uwaga: Produkty te obejmują między innymi pewne związki powierzchniowo czynne oraz zawierające je preparaty oparte np. na olejach, estrach kwasu fosforowego i wielkocząsteczkowych alkoholach oraz produktach ich oksyalkilenowania.

40 anti-perspirant
fr contre la transpiration
de schweisshemmend
el antiperspirante
it contro la traspirazione
ne anti-zweetmiddel
pl przeciwpotowy

41 anti-redepositing power
The ability of a substance to prevent insoluble particles from redepositing on the washed surface and, possibly, to maintain the particles in suspension.

fr pouvoir d'antiredéposition
Degré d'aptitude d'un produit à empêcher les particules insolubles de se déposer à nouveau sur la surface lavée, et à maintenir éventuellement ces particules en suspension.

de Schmutztragevermögen
Grad der Fähigkeit eines Produktes, unlösliche Teilchen in Suspension zu halten und/oder sie zu hindern, sich auf der gereinigten Oberfläche wieder abzulagern.

el poder de anti-redeposición
Capacidad de un producto para mantener en suspensión partículas insolubles y/o para impedir que se depositen de nuevo sobre la superficie lavada.

it potere di antirideposizione
Grado dell'attitudine di una sostanza ad impedire che particelle insolubili si ridepositano sulla superficie lavata ed,

eventualmente, a mantenere tali particelle in sospensione.

ne vuildragend vermogen, antivergrauwingsvermogen
Het vermogen van een stof om te voorkomen, dat onoplosbare deeltjes zich weer op het gewassen oppervlak afzetten en om zo nodig de deeltjes in gesuspendeerde vorm te houden.

pl zdolność ochronna przed wtórnym osadzaniem się brudu
Zdolność danej substancji do przeciwdziałania osadzaniu się na pranej powierzchni nierozpuszczalnych cząstek występujących w zawiesinie, jak również do ewentualnego utrzymywania tych cząstek w zawiesinie.

42 anti-redeposition agent
A complementary component of a detergent, usually organic, which imparts to the latter the property of preventing redeposition. Example: carboxymethylcellulose.

fr agent d'antiredéposition
Composant complémentaire d'un détergent, généralement organique, qui lui confère un pouvoir d'antiredéposition. Exemple: carboxyméthylcellulose.

de Schmutzträger
Zusätzlicher Bestandteil eines Waschmittels, im allgemeinen organisch, welcher ihm Schmutztragevermögen verleiht. Beispiele: Carboxymethylcellulose, etc.

el agente de anti-redeposición
Componente complementario de un detergente, de naturaleza generalmente orgánica, que le confiere un poder de anti-redeposición. Ejemplos: carboximetilcelulosa, etc.

it agente di antirideposizione
Costituente complementare di un detergente, generalmente organico, che gli conferisce il potere di antirideposizione. Ad es. carbossimetilcellulosa.

ne antiredepositie-middel, antivergrauwingsmiddel
Een complementair bestanddeel van een wasmiddel, gewoonlijk organisch, dat hieraan de eigenschap verleent het weer neerslaan van het vuil te voorkomen, en om de deeltjes in gesuspendeerde vorm te houden. Voorbeeld: carboxymethylcellulose.

pl środek zapobiegający wtórnemu osadzaniu się brudu (pigmentowego)
Dodatek, na ogół organiczny, do środków piorących nadający im zdolność przeciwdziałania wtórnemu osadzaniu się brudu. Przykład: karboksymetyloceluloza.

43 antiseptic hand cream
fr crème antiseptique pour les mains
de antiseptische Handcreme
el crema antiséptica para las manos
it crema antisettica per le mani
ne antiseptische handcrème
pl antyseptyczny krem do rąk

44 anti-wrinkle cream
fr crème anti-rides
de hautstraffende Creme
el crema contra las arrugas
it crema antirughe
ne crème tegen rimpels
pl krem przeciw zmarszczkom

45 apparent density
Mass of the unit of apparent volume.
fr masse volumique apparente
Masse de l'unité de volume apparent.
de scheinbare spezifische Masse
Masse der Einheit des scheinbaren Volumens.
el masa específica aparente
Masa de la unidad de volumen aparente.
it —
ne schijnbaar soortelijk gewicht
Gewicht van de eenheid van het schijnbaar volume.
pl pozorny ciężar właściwy
Masa jednostki pozornej objętości.

46 apparent volume
Volume determined by the exterior limits of a quantity of substance, under the experimental test conditions. This volume includes possible bubbles, pores and interstices.
fr volume apparent
Volume déterminé par les limites extérieures d'une quantité de substance, dans les conditions expérimentales de l'essai. Ce volume comprend les bulles, pores et interstices éventuels.
de scheinbares Volumen
Volumen einer durch äussere Grenzen fixierten Substanzmenge unter den ex-

perimentellen Bedingungen. Dieses Volumen umfasst eventuelle Blasen, Poren und Zwischenräume.

el volumen aparente
Volumen determinado por los límites exteriores de una cantidad de substancia, en las condiciones experimentales del ensayo. Este volumen comprende las burbujas, poros e intersticios que puedan existir dentro de dichos límites.

it —

ne schijnbaar volume
Volume bepaald door de buitenste begrenzing van een hoeveelheid substantie onder de omstandigheden van de proef. In dit volume zijn begrepen eventuele blaasjes, belletjes, poriën en tussenruimten.

pl objętość pozorna
Objętość określona w warunkach doświadczalnych przez zewnętrzne granice danej ilości substancji. W objętości tej zawarte mogą być ewentualne pęcherzyki, pory i szczeliny.

47 application technique
fr technique d'application
de Anwendungstechnik
el técnica de aplicación
it tecnica di applicazione
ne toepassingstechniek
pl sposób użycia, sposób zastosowania

48 aqueous emulsion (symbol L-H: oil in water[1])
An emulsion in which the continuous phase is aqueous.

fr émulsion de type aqueux (symbole L-H: huile dans l'eau[2])
Émulsion dont la phase continue est aqueuse.

de wässrige Emulsion (Abkürzung L-H = Öl in Wasser[3])
Emulsion, deren kontinuierliche Phase wässrig ist.

el emulsión acuosa
Emulsión, cuya fase continua es acuosa.

it emulsione di tipo acquoso
Emulsione la cui fase continua è acquosa.

ne waterige emulsie (symbool L-H: olie in water)
Een emulsie waarin water de continue fase vormt.

pl emulsja wodna (symbol L-H: olej w wodzie[4])

Emulsja, w której fazą rozpraszającą jest woda.
[1] From the Greek: L = lipos, H = hydor.
[2] Du grec: L = lipos, H = hydor.
[3] Vom griechischen: L = lipos, H = hydor.
[4] Z greckiego: L = lipos, H = hydor.

49 artificial soil
Soil of selected composition, prepared for detergency tests.
fr salissure artificielle
Salissure de composition choisie, préparée en vue d'essais relatifs à la détergence.
de künstliche Anschmutzung
Anschmutzung ausgewählter Zusammensetzung, hergestellt zur Durchführung von Waschversuchen jeder Art.
el suciedad artificial
Suciedad de composición seleccionada para ensayos relativos a la detergencia.
it sporco artificiale
Sporco di composizione determinata, preparato per le prove di detergenza.
ne kunstmatig vuil
Vuil van gekozen samenstelling, bereid voor wasproeven.
pl zabrudzenie sztuczne
Zabrudzenie o odpowiednio dobranym składzie, sporządzane do badania zdolności piorącej.

50 artificially soiled test cloth
fr tissu artificiellement sali, tissu expérimental sali
de künstlich angeschmutztes Testgewebe
el tejido ensuciado artificialmente
it tessuto artificialmente sporcato
ne kunstmatig bevuilde proefdoek
pl tkanina testowa sztucznie zabrudzona

51 ash content
fr teneur en cendres, taux des cendres
de Aschegehalt
el contenido en cenizas, porcentaje de cenizas
it contenuto in ceneri
ne asgehalte
pl zawartość popiołu

52 atomisation
fr pulvérisation
de Zerstäubung
el pulverización, atomización
it atomizzazione
ne verstuiven
pl rozpylanie

53 autoxidation
Chemical reaction involving the unaided coupling, either fast or slow, of molecular oxygen to an organic or inorganic compound.

fr autoxydation
Réaction chimique non induite consistant en la fixation plus ou moins rapide de l'oxygène moléculaire sur une substance chimique organique ou inorganique.

de Autoxydation
Nicht induzierte chemische Reaktion, bei der molekularer Sauerstoff mehr oder weniger rasch von einer organischen oder anorganischen Verbindung chemisch gebunden wird.

el autoxidación
Reacción química no inducida que consiste en la fijación, más o menos rápida, de oxígeno molecular en una substancia química orgánica o inorgánica.

it autossidazione
Reazione chimica spontanea che consiste nella fissazione più o meno rapida di ossigeno molecolare su una sostanza chimica organica o inorganica.

ne autoxidatie
Chemische reactie waarbij moleculaire zuurstof snel of langzaam spontaan wordt geaddeerd aan een organische of anorganische verbinding.

pl samoutlenianie, utlenianie samorzutne
Nieindukowana reakcja chemiczna szybkiego lub powolnego przyłączania tlenu cząsteczkowego do związków organicznych lub nieorganicznych.

54 available oxygen
fr oxygène actif
de Aktivsauerstoff
el oxígeno activo
it ossigeno attivo
ne aktieve zuurstof
pl aktywny tlen

B

55 baby cream
fr crème pour bébés
de Baby-Creme
el crema para bebes
it crema per bambini
ne babycrème
pl krem dla niemowląt

56 bactericidal action, antibacterial action
fr action bactéricide, action antibactérienne
de bakterizide Wirkung
el acción bactericida, acción antibacteriana
it azione antibacterica
ne bactericide werking
pl działanie bakteriobójcze, działanie przeciwbakteryjne

57 baldness
fr calvitie
de Kahlheit
el calvicie
it calvizie
ne kaalheid
pl łysina

58 ballpoint applicator
fr applicateur à bille
de Rollstift
el aplicador de punta redonda
it applicatore a sfera
ne bolpen, roller
pl aplikator kulkowy

59 bar soap
fr savon en barres, savonnette, pain de savon
de Stückseife
el jabón en barra, jaboncillo
it sapone in pezzi, saponette, pezzo di sapone
ne stukzeep
pl mydło w kawałku, mydło w sztabach

60 base coat
fr base pour les ongles, vernis base
de grundierender Überzug
el base para las uñas, barniz base
it base per le unghie
ne grondlaag
pl podkład (pod lakier do paznokci)

61 bath ratio, loading ratio
fr rapport de bain
de Flottenverhältnis
el relación del baño
it rapporto di bagno
ne vlotverhouding
pl moduł kąpieli, stosunek kąpieli

62 bath soap
fr savonnettes pour le bain
de Badeseife
el jabón de baño
it saponetta per bagno
ne badzeep
pl mydło kąpielowe

63 beads (hollow)
fr sphérules creuses, billes creuses
de hohlkugelförmige Körnchen
el gránulos esféricos huecos
it sferette cave
ne holle korrels, holle bolletjes
pl kulki, granulki proszku do prania (wewnątrz puste)

64 beauty mask
fr masque de beauté, masque facial
de Gesichtspackung, Schönheitspackung
el máscara de belleza, máscara facial
it maschera di bellezza, maschera facciale
ne schoonheidsmasker
pl maseczka upiększająca

65 beta phase
The crystalline form, stable in the cold, which is produced by the slow cooling of the finished soap to below a critical temperature or by mechanical action (kneading, milling, compression, extrusion or shearing) exerted in the cold to the omega phase. It is the predominant phase in soaps processed mechanically. Soaps of low molecular weight (coconut oil, palm-kernel oil etc.) are not converted into beta phase or only form it very slowly. Soaps in beta phase have a high solution rate and, in consequence, foam easily. They are firmer than soaps in the delta or omega phases.
fr phase bêta
Forme cristalline, stable à froid, qui se produit par refroidissement lent du savon lisse au-dessous d'une température critique, ou par action mécanique (pétrissage, laminage, compression, extrusion ou cisaillement) exercée sur la phase oméga à froid. C'est la phase

prédominante dans les savons traités mécaniquement. Les savons à bas poids moléculaire (savon d'huile de coprah, palmiste, etc.) ne se transforment pas en phase bêta ou le font très lentement. Les savons en phase bêta possèdent une vitesse de dissolution élevée et en conséquence moussent facilement. Ils sont plus fermes que les savons en phase delta ou oméga.

de Beta-Phase
Kristalline in der Kälte stabile Form, welche durch langsame Abkühlung des geschliffenen Seifenkerns unterhalb einer kritischen Temperatur oder durch mechanische Bearbeitung (Kneten, Walzen, Strangpressen oder Scheren) einer kalten Seife in der Omega-Form, entsteht. Diese Phase ist in den pilierten Seifen vorzugsweise enthalten. Die Seifen aus Fettsäuren mit niedrigem Molekulargewicht (Kokos- bzw. Palmkernölfettsäure) gehen nicht oder nur langsam in die Beta-Phase über. Seifen, die in Form der Beta-Phase vorliegen, besitzen eine erhöhte Lösungsgeschwindigkeit und schäumen daher rasch an. Sie sind härter als die Seifen der Delta- oder Omega-Phase.

el fase beta
Forma cristalina, estable en frío, que se forma por enfriamiento lento del jabón liquidado por debajo de una temperatura crítica (unos 70°C), o por una acción mecánica ejercida sobre el jabón frío en fase omega, como por ejemplo, por las fuerzas de cizallamiento que se producen por amasado, laminación o compresión. Es la fase predominante en los jabones maquinados. Los jabones de ácidos grasos de bajo peso molecular (jabones de aceite de coco y palmiste) no se transforman en fase beta o lo hacen muy lentamente. Los jabones en fase beta poseen una velocidad de disolución elevada y como consecuencia producen fácilmente espuma. Son además más duros que los jabones en fase delta u omega.

it fase beta
Forma cristallina, stabile a freddo, che è prodotta da lento raffreddamento del sapone finito, al disotto della temperatura critica, o per azione meccanica (impasto, macinazione con pressione, estrusione o triturazione) effettuata nella fase omega a freddo. E' la fase predominante nella produzione del sapone per via meccanica. I saponi di basso peso molecolare, di olio di cocco, olio di palmisto, ecc. non sono convertiti nella fase beta, o solo in minima parte. I saponi di fase beta hanno un'alta solubilità, e di conseguenza producono facilmente schiuma. Essi sono più compatti dei saponi della fase delta o omega.

ne beta-fase
Een bij lagere temperatuur stabiele kristallijnen vorm, die door langzame afkoeling van de geslepen zeepkern onder een kritische temperatuur of door mechanische bewerking (kneden, walsen, stangpersen of afschuiven) van een koude zeep in de omega-vorm ontstaat. Dit is de voornaamste fase in mechanisch bewerkte zepen. Zepen met laag moleculair-gewicht (kokosnootolie, palmkernolie enz.) worden niet in beta-fase omgezet of vormen het slechts heel langzaam.

pl faza beta
Krystaliczna forma mydła, trwała na zimno, utworzona bądź podczas powolnego schładzania wysołu płynnego (rdzenia) poniżej temperatury krytycznej, bądź przez mechaniczną obróbkę na zimno mydła w fazie omega (ugniatanie, walcowanie, wyciskanie lub ścinanie). Jest to główna faza w mydle poddanym obróbce mechanicznej. Mydła otrzymane z niskocząsteczkowych kwasów tłuszczowych (np. z oleju kokosowego lub z oleju z ziarn palmowych) nie przechodzą w ogóle, bądź przechodzą bardzo powoli w fazę beta. Mydła w fazie beta charakteryzują się zwiększoną szybkością rozpuszczania, w wyniku czego łatwo się pienią. Są one twardsze od mydeł w fazach delta lub omega.

66 binding agent, binder
fr agent liant, épaississant, liant, gonflant
de Bindemittel
el agente espesante, aglutinante
it agente legante
ne bindmiddel
pl środek wiążący, spoiwo

67 biodegradability
Aptitude of an organic substance to undergo biodegradation.
fr biodégradabilité
Aptitude d'une matière organique à subir la biodégradation.
de biologische Abbaubarkeit
Fähigkeit einer organischen Substanz, dem biologischen Abbau zu unterliegen.
el biodegradabilidad
Aptitud de una materia orgánica para experimentar una biodegradación.
it biodegradabilità
Attitudine di una sostanza organica alla biodegradazione.
ne biologische afbreekbaarheid, biodegradatie
Vermogen van een organische stof om biologisch afgebroken te worden.
pl rozkładalność biologiczna, biodegradacja
Zdolność substancji organicznych do ulegania rozkładowi biologicznemu.

68 biodegradable surface active agent
Surface active agent which is susceptible to biodegradation causing a loss of its surface active characteristics.
fr agent de surface biodégradable
Agent de surface qui est susceptible d'une biodégradation conduisant à une perte de ses caractéristiques d'activité de surface.
de biologisch abbaubares Tensid
Tensid, welches einem biologischen Abbau unterworfen werden kann und dadurch seine Grenzflächenaktivität verliert.
el agente de superficie biodegradable
Agente de superficie que puede experimentar una biodegradación, que conduzca a una pérdida de sus características de actividad de superficie.
it tensioattivo biodegradabile
Tensioattivo che è suscettibile di biodegradazione in presenza di acqua, con conseguente perdita della sua tensioattività.
ne biologisch afbreekbare oppervlakaktieve stof
Oppervlakaktieve stof die een verandering kan ondergaan door biodegradatie waardoor zijn oppervlakaktieve eigenschappen verloren gaan.
pl rozkładalny biologicznie związek powierzchniowo czynny

Związek powierzchniowo czynny ulegający rozkładowi biologicznemu, w wyniku którego traci on swoje własności.

69 biodegradation
Molecular degradation of an organic substance ordinarily in an aqueous medium resulting from the complex actions of living organisms.
fr biodégradation
Dégradation moléculaire d'une matière organique en milieu généralement aqueux résultant des actions complexes d'organismes vivants.
de biologischer Abbau
Molekularer Abbau einer organischen Substanz in einem (im allgemeinen) wässrigen Milieu, welcher durch komplexe Einwirkungen lebender Organismen hervorgerufen wird.
el biodegradación
Degradación molecular de una materia orgánica, en medio generalmente acuoso, como resultado de acciones complejas de organismos vivos.
it biodegradazione
Demolizione di una struttura molecolare dovuta all'azione di microrganismi viventi.
ne biologische afbraak
Moleculaire afbraak van een organische stof die het gevolg is van de komplexe aktiviteit van levende organismen.
pl rozkład biologiczny, biodegradacja
Rozkład cząsteczek substancji organicznej zachodzący zazwyczaj w środowisku wodnym w wyniku złożonego działania żywych organizmów.

70 bleach, to
fr décolorer
de bleichen
el decolorar
it decolorare
ne bleken
pl rozjaśniać, bielić, wybielać

71 bleach cream
fr crème à blanchir
de Bleichcreme
el crema para blanquear
it crema imbiancante
ne bleekcrème
pl krem rozjaśniający, krem wybielający

72 bleaching
fr décoloration

de Bleichen
el decoloración
it decolorazione
ne bleking
pl rozjaśnianie, bielenie, wybielanie

73 **bleaching agent, bleaching product**
fr préparation décolorante, produit décolorant
de Bleichmittel
el preparación decolorante
it preparazione decolorante, prodotto decolorante
ne bleekmiddel
pl środek rozjaśniający, środek bielący, środek wybielający

74 **bleaching assistant**
Product which makes it possible to speed up the bleaching operation and to make it more even in effect.
Note: These products are generally surface active agents, mainly wetting agents which are stable in bleaching baths.

fr adjuvant de blanchiment
Produit qui permet d'accélérer l'opération de blanchiment et de lui assurer un effet plus régulier.
Nota: Il s'agit en général d'agents de surface, principalement agents mouillants, stables dans les bains de blanchiment.

de Bleichhilfsmittel
Produkt, das es ermöglicht, den Bleichprozess schneller durchzuführen und den Bleicheffekt gleichmässiger zu gestalten.
Anmerkung: Es handelt sich im allgemeinen um Erzeugnisse auf der Basis von grenzflächenaktiven Stoffen, d.h. um gegen Bleichflotten beständige Netzmittel.

el auxiliar de blanqueo
Producto que permite acelerar la operación de blanqueo y obtener un efecto más regular.
Observación: Se trata, en general, de productos a base de agentes de superficie, es decir, de agentes humectantes estables en los baños de blanqueo.

it —
ne bleekhulpmiddel
Produkt hetwelk tot doel heeft de effectiviteit en snelheid van het bleekproces te verbeteren.

pl środek pomocniczy do rozjaśniania
Produkt umożliwiający przyśpieszanie operacji rozjaśniania oraz otrzymanie bardziej równomiernego efektu.
Uwaga: Produktami takimi są na ogół związki powierzchniowo czynne, głównie związki zwilżające trwałe w kąpielach bielących.

75 **bleaching efficiency**
fr pouvoir blanchissant
de Bleichwirkung
el poder blanqueante
it potere sbiancante, potere candeggiante
ne bleekwerking
pl zdolność rozjaśniania, zdolność bieląca

76 **bleaching liquor, bleaching bath**
fr bain de blanchiment
de Bleichflotte
el baño de blanqueo, líquido de blanqueo
it bagno di sbianca, bagno di candeggio
ne bleekbad
pl kąpiel rozjaśniająca, kąpiel bieląca, roztwór rozjaśniający

77 **bleaching process**
fr opération de blanchiment
de Bleichprozess
el proceso de blanqueo, operación de blanqueo
it operazione di sbianca, operazione di candeggio
ne bleekproces
pl proces rozjaśniania, proces bielenia

78 **boiling**
The operation aimed at completing the saponification reaction by boiling the soap mass with excess alkali hydroxide, either in isotropic solution or in a heterogeneous system "curd soap–lye" (Marseilles process).

fr cuisson (coction)
Opération ayant pour but de mener à son terme la réaction de saponification, par ébullition de la masse savonneuse avec de l'hydroxyde alcalin en excès, soit en solution isotrope, soit en système hétérogène "savon grainé–lessive inférieure" (procédé marseillais).

de Sieden (Klarsieden)
Prozess des Aufkochens der Seifenmasse mit überschüssigen Alkalien zur Vollendung der Verseifung. Das Sieden wird in isotroper Lösung oder im heterogenen System "geronnener Kern-Unterlauge" durchgeführt.

el cocción
Ebullición de la masa de jabón con un exceso de álcali, para completar la saponificación. La cocción tiene lugar en general en disolución isótropa.

it bollitura
L'operazione mira al completamento

della reazione di saponificazione, effet-
tuando la bollitura della massa di
sapone con un eccesso di alcali caustico,
sia nel caso di soluzioni isotrope, che
nel caso di sistemi eterogenei "sapone
levato–liscivia alcalina" (Sapone Marsiglia).

ne zieden
Proces om door opkoken van de zeepmassa met overmaat alkaliën de verzeping te voleindigen. Het zieden wordt
in isotrope oplossing of in het hetero-
gene systeem "kernzeep–onderloog"
uitgevoerd.

pl domydlanie osnowy
Operacja mająca na celu zakończenie
reakcji zmydlania przy warzeniu masy
mydlanej, polegająca na gotowaniu jej
z nadmiarem alkaliów bądź w roztworze
izotropowym, bądź w układzie hete-
rogenicznym "wysół ścięty–ług spodni".

79 booster (for surface active agents)
A complementary component of a detergent, usually organic, which strength-
ens certain characteristic properties
of the essential components. Examples:
certain alkylolamides and amine oxides.

fr renforçateur (pour agents de surface)
Composant complémentaire d'un détergent, généralement organique, qui
améliore certaines propriétés caractéristiques des composants essentiels.
Exemples: certains alkylolamides et
oxydes d'amines.

de Verstärker (für Tenside)
Zusätzlicher, im allgemeinen organischer
Bestandteil eines Waschmittels, welcher
gewisse charakteristische Eigenschaften
der Hauptbestandteile verbessert. Beispiele: gewisse Alkanolamide, Amin-
oxide, etc.

el reforzador (para agentes de superficie)
Componente complementario y generalmente orgánico de un detergente,
que mejora alguna de las propiedades
características de los componentes esenciales. Ejemplos: ciertas alcanolamidas,
ciertos óxidos de aminas, etc.

it rafforzatore
Costituente complementare di un detergente, generalmente organico, che
esalta certe caratteristiche proprietà
dei costituenti principali. Esempi:
alcanolamidi ecc.

ne booster (voor oppervlakaktieve stoffen)
Complementair bestanddeel van een
wasmiddel, gewoonlijk organisch, waardoor bepaalde kenmerkende eigenschappen van de hoofdbestanddelen
versterkt worden. Voorbeelden: sommige
alkanolamiden en amino-oxiden.

pl wspomagacz (dla związków powierzchniowo czynnych)
Dodatkowy, na ogół organiczny składnik środków piorących, poprawiający
pewne charakterystyczne własności składników podstawowych. Przykłady: niektóre alkanoloamidy oraz tlenki amin.

80 breath freshener
fr rafraîchissant pour l'haleine
de Mittel zur Erfrischung des Atems
el refrescante para el aliento
it rinfrescante dell'alito
ne ademzuiveringsmiddel
pl dezodorant do ust

81 brightening agent
Product intended to improve or restore
the purity of the colour; it may also
play a part in giving the textile articles
the desired finish from the point of
view of feel and sheen.
Note: These products are generally those mentioned under the heading of preparing agent.

fr agent d'avivage
Produit destiné à améliorer ou à resti-
tuer la pureté du coloris; il peut contri-
buer également à donner aux articles
textiles l'effet d'apprêt désiré du point
de vue toucher et brillant.
Nota: Il s'agit généralement de produits cités sous
agent de préparation.

de Avivagemittel
Produkt, das die Reinheit der Färbung
verbessert oder wiederherstellt. Es gibt
textilen Erzeugnissen aber auch hinsicht-
lich Griff und Glanz den gewünschten
Ausrüstungseffekt.
Anmerkung: Es handelt sich in der Regel um
Produkte, die bereits unter Präparationsmittel
genannt sind.

el agente de avivado
Producto destinado a mejorar o resti-
tuir la pureza del colorido. Contribuye
igualmente a dar a los artículos textiles
un elevado brillo y un tacto suave.
Observación: Se trata generalmente de los productos citados en agente de preparación.

it —
ne optisch bleekmiddel
Een produkt dat tot taak heeft de zui-

verheid van kleur te verbeteren of te herstellen. Het kan ook een funktie hebben bij het geven van de gewenste finish aan textiel met betrekking tot greep en glans.

Opmerking: Gewoonlijk zijn dit de produkten genoemd onder preparing agent.

pl środek ożywiający (rozjaśniający) barwę Produkt, którego zadaniem jest poprawa lub przywrócenie czystości barwy. Produkt ten może odgrywać także pewną rolę w nadawaniu wyrobom włókienniczym pożądanego wykończenia pod względem dotyku (chwytu) i połysku.

Uwaga: Produkty te, ogólnie biorąc, objęte są definicją środków do preparacji.

82 brightening effect
fr éclaircissement
de Aufhellung
el efecto de brillo
it nuanza (sfumatura)
ne tintverbetering
pl rozjaśnienie

83 brushless shaving cream
fr crème sans blaireau, crème rapide
de nichtschäumende Rasiercreme
el crema de afeitar sin brocha, crema rápida
it crema da barba senza pennello, crema rapida
ne niet schuimende scheercrème
pl krem do golenia bezpędzlowy

84 bubble
A volume of gas enclosed by a thin film of liquid.
fr bulle
Volume gazeux limité par une enveloppe liquide mince.
de Gasblase
Mit Gas gefüllter Hohlraum, der von einer dünnen flüssigen Hülle begrenzt ist.
el burbuja
Volumen gaseoso, limitado por una envoltura líquida delgada.
it bolla
Volume di gas delimitato da uno strato sottile di liquido.
ne bel
Een met gas gevulde ruimte omsloten door een dunne vloeistoffilm.
pl pęcherzyk
Przestrzeń wypełniona gazem i ograniczona cienką błonką cieczy.

85 buffer capacity of the skin
fr capacité-tampon de la peau
de Pufferkapazität der Haut
el capacidad tope de la piel, capacidad tampón de la piel
it potere tampone della pelle
ne buffer-capaciteit van de huid
pl pojemność buforująca skóry

86 builder (for surface active agents)
A subsidiary constituent of a detergent, usually inorganic, which, with reference to the washing action, adds its characteristic properties to those of the essential components. Examples: polyphosphates, carbonates, silicates, etc.
fr adjuvant (pour agents de surface)
Composant complémentaire d'un détergent, souvent minéral, qui ajoute ses propriétés particulières à celles des composants essentiels, quant à l'action spécifique du lavage. Exemples: polyphosphates, carbonates, silicates, etc.
de Builder (für Tenside)
Zusätzlicher, im allgemeinen anorganischer Bestandteil eines Waschmittels, welcher in Bezug auf den Waschprozess seine speziellen Eigenschaften denen des Waschmittels hinzufügt. Beispiele: Polyphosphate, Karbonate, Silikate, usw.
el coadyuvante (para agentes de superficie)
Componente complementario y generalmente inorgánico de un detergente, que, por lo que respecta a la acción específica del lavado, añade sus propiedades particulares a las de los componentes esenciales. Ejemplos: polifosfatos, carbonatos, silicatos, etc.
it adiuvante
Costituente complementare di un detergente, generalmente inorganico, che per quanto riguarda l'azione specifica del lavaggio aggiunge le sue proprietà caratteristiche a quelle dei costituenti principali. Esempi: polifosfati, carbonati, silicati ecc.
ne builder (voor oppervlakaktieve stoffen)
Een complementair bestanddeel van een wasmiddel, gewoonlijk anorganisch, dat met betrekking tot de waswerking zijn kenmerkende eigenschappen voegt bij die van de hoofdbestanddelen. Voorbeelden: polyfosfaten, carbonaten, silicaten.

pl wypełniacz (dla związków powierzchniowo czynnych)
Dodatkowy na ogół nieorganiczny składnik środków piorących, który w odniesieniu do procesu prania uzupełnia swymi charakterystycznymi własnościami własności składników podstawowych. Przykłady: polifosforany, węglany, krzemiany itd.

87 **bulk density**
fr poids spécifique apparent
de Schüttgewicht
el peso específico aparente
it densità apparente, peso specifico apparente
ne schijnbaar soortelijk gewicht
pl ciężar objętościowy, ciężar nasypowy

C

88 caking
fr encollement, agglutination
de Balligwerden, Zusammenbacken, Zusammenballung
el aglutinación
it impaccamento
ne bakken, samenbakken, samenkoeken
pl zbrylanie

89 calcification of teeth
fr calcification des dents
de Kalzifizierung der Zähne
el calcificación de los dientes
it calcificazione dei denti
ne verkalking der tanden
pl uwapnienie zębów

90 calcium chelating power, calcium sequestering power
fr pouvoir anticalcaire
de Kalkbindevermögen, Calciumkomplexbildungsvermögen
el poder anticalcáreo, poder secuestrante calcio
it potere sequestrante per il calcio
ne kalkbindend vermogen
pl zdolność chelatowania wapnia, zdolność sekwestrowania wapnia, zdolność kompleksowania wapnia

91 calcium hardness
fr dureté calcaire, dureté calcique
de Kalkhärte
el dureza de calcio, dureza cálcica
it durezza calcarea
ne kalkhardheid
pl twardość wapniowa

92 capillary activity
The property of a substance in solution to decrease interfacial tension, associated with an augmentation of the concentration of the substance at the interface.
fr tensio-activité
Propriété d'un produit en solution d'abaisser la tension interfaciale, associée à une augmentation de la concentration du produit à l'interface.
de Kapillaraktivität
Fähigkeit einer Verbindung in gelöstem Zustand die Oberflächenspannung herabzusetzen, bei gleichzeitiger Konzentrationserhöhung in der Oberfläche.

el tensioactividad
Propiedad de una substancia en disolución de rebajar la tensión interfacial, asociada a un aumento de su concentración en la interfacie.
it tensioattività
Proprietà di una sostanza in soluzione di diminuire la tensione interfacciale, legata ad un aumento della sua concentrazione all'interfaccia.
ne capillaire aktiviteit
De eigenschap van een stof in oplossing, de grensvlakspanning te verlagen, gecombineerd met een verhoging van de concentratie van de stof aan het grensvlak.
pl aktywność powierzchniowa
Zdolność substancji w roztworze do obniżania napięcia międzyfazowego, związana ze wzrostem jej stężenia na granicy faz.

93 carbonizing assistant
Product facilitating and speeding up the penetration of carbonizing agents (acids or generators of acids) into the vegetable impurities in wool, thus promoting their destruction during subsequent heat treatment.
Note: These products are wetting agents with sufficient stability in the presence of acids.
fr adjuvant de carbonisage
Produit facilitant et accélérant la pénétration des agents de carbonisage (acides ou générateurs d'acides) dans les impuretés végétales de la laine, favorisant ainsi leur destruction au cours du traitement thermique subséquent.
Nota: Il s'agit d'agents mouillants à stabilité suffisante envers les acides.
de Karbonisierhilfsmittel
Produkt, das das Eindringen der Karbonisiermittel (Säuren oder säureabspaltende Verbindungen) in die vegetabilischen Verunreinigungen der Wolle erleichtert und beschleunigt, wodurch eine Zerstörung der pflanzlichen Verunreinigungen bei der anschliessenden thermischen Behandlung begünstigt wird.
Anmerkung: Es handelt sich um ausreichend säurebeständige Netzmittel.
el auxiliar de carbonizado
Producto que facilita y acelera la penetración de los agentes de carbonizado

(ácidos o generadores de ácidos) en las impurezas vegetales de la lana, favoreciendo así su destrucción en el curso del tratamiento térmico subsiguiente.

Observación: Se trata de agentes humectantes con suficiente estabilidad para los ácidos.

it —

ne carboniseerhulpmiddel
Produkt dat de penetratie van carboniseermiddelen (zuren en zuurvormers) in de verontreinigingen van wol vergemakkelijkt en versnelt, zodat hun afbraak tijdens de volgende warmtebehandelingen versneld wordt.

Opmerking: Dit zijn bevochtigingsmiddelen met een voldoende stabiliteit in aanwezigheid van zuren.

pl środek wspomagający karbonizację, wspomagacz karbonizacji
Produkt ułatwiający i przyśpieszający penetrację środków karbonizujących (kwasy lub związki wydzielające kwasy) w głąb zanieczyszczeń wełny pochodzenia roślinnego, pobudzający ich rozkład w czasie prowadzonej następnie obróbki cieplnej.

Uwaga: Produktami takimi są związki zwilżające o dostatecznej trwałości w obecności kwasów.

94 carboxymethylcellulose (CMC)
fr carboxyméthylcellulose
de Carboxymethylcellulose, celluloseglykolsaures Natrium
el carboximetilcelulosa
it carbossimetilcellulosa
ne carboxymethylcellulose (CMC)
pl karboksymetyloceluloza

95 carious spot
fr point de carie
de kariöse Stelle
el mancha de caries
it punto di carie
ne carieuze plek
pl plama próchnicza

96 cationic surface active agent
A surface active agent which has one or more functional groups and ionizes in aqueous solution to produce positively charged organic ions responsible for the surface activity.
fr agent de surface cationique
Agent de surface possédant un ou plusieurs groupements fonctionnels s'ionisant en solution aqueuse, pour fournir des ions organiques chargés positive-

ment et responsables de l'activité de surface.
de kationische grenzflächenaktive Verbindung
Grenzflächenaktive Verbindung, die eine oder mehrere funktionelle Gruppen besitzt, die in wässriger Lösung ionisieren unter Bildung positiv geladener organischer Ionen, die für die Grenzflächenaktivität verantwortlich sind.
el agente de superficie catiónico
Agente de superficie que posee uno o varios grupos funcionales, que se ionizan en disolución acuosa, dando iones orgánicos cargados positivamente, responsables de la actividad de superficie.
it tensioattivo cationico
Tensioattivo con uno o più gruppi funzionali, che si ionizzano in soluzione acquosa formando ioni organici carichi positivamente e responsabili della tensioattività.
ne kationaktieve stof
Een oppervlakaktieve stof die één of meer funktionele groepen bevat en die in een waterige oplossing een ioniserende werking heeft en zodoende positief geladen organische ionen produceert, die de oppervlakaktiviteit teweegbrengen.
pl kationowy związek powierzchniowo czynny
Związek powierzchniowo czynny zawierający jedną lub więcej grup funkcyjnych, ulegający jonizacji w roztworze wodnym z utworzeniem dodatnio naładowanych jonów odpowiedzialnych za aktywność powierzchniową.

97 cationics
fr composés cationiques
de kationaktive Substanzen
el compuestos catiónicos
it composti cationici
ne kationaktieve stoffen
pl kationowe związki powierzchniowo czynne

98 centrifugal disk atomisation
fr atomisation à disque tournant
de Zerstäubung mittels rotierenden Tellers
el atomización por disco rotativo, atomización por disco giratorio
it atomizzazione a disco rotante
ne verstuiving door middel van een draaiende schijf
pl rozpylanie tarczą wirującą

99 centrifugal disk atomiser, spinning disk atomiser, centrifugal disk spray wheel
fr atomiseur à disque tournant, disque atomiseur
de Fliehkraftzerstäuber, Zerstäuberscheibe
el atomizador de disco rotativo, disco atomizador
it atomizzatore a disco rotante, disco rotante
ne draaischijfverstuiver
pl rozpylacz z wirującą tarczą

100 change in shade
fr virage de nuance
de Verschiebung des Farbtones, Verschiebung des Farbstichs
el cambio del matiz, viraje del matiz
it viraggio della nuanza
ne verschuiving van de tint
pl zmiana odcienia

101 chelate
A metallic complex in which the metal ion is suppressed by chelation.
fr chélate
Complexe métallique où l'ion métal est dissimulé par chélation.
de Chelat
Durch Chelatbildung entstandener, ringförmiger Metallkomplex.
el quelato
Complejo metálico en el que el ion metal se encuentra enmascarado por quelación.
it chelato
Complesso dove l'ione metallico è dissimulato per chelazione.
ne metaalcomplex
Een complex waarin de ionisatie van het metaalion door een complexvorming wordt teruggedrongen.
pl chelat
Kompleks metalu utworzony w wyniku reakcji chelatowania (tworzenia pierścieni chelatowych).

102 chelating agent
A substance having a molecular structure embodying several electron-donor groups which render it capable of combining with metallic ions by chelation.
fr agent chélatant, chélatant
Produit dont la molécule comporte plusieurs groupements donneurs d'électrons, le rendant apte à engager les ions métalliques en chélation.

de Chelatbildner
Produkt, dessen Molekül eine oder mehrere Elektronen abgebende Gruppen besitzt, durch welche es befähigt ist, Metallionen in einem ringförmigen Komplex zu binden.
el agente quelante, quelatante
Producto cuya molécula posee varios grupos donantes de electrones, gracias a los cuales puede fijar iones metálicos por quelación.
it chelante
Sostanza la cui molecola comprende uno o più gruppi donatori di elettroni idonea a legare gli ioni metallici in chelazione.
ne complexvormer
Een stof met een molecuulstructuur waarin zich een aantal elektron-donorgroepen bevindt dat de stof in staat stelt zich door complexvorming met metaalionen te verbinden.
pl środek chelatujący
Substancja zawierająca w cząsteczce szereg grup będących donorami elektronów. Grupy te sprawiają, że substancja ta jest zdolna do wiązania jonów metali w reakcji chelatowania.

103 chelating power
The ability of certain bodies to complex cations to form a so-called ring structure.
fr pouvoir de chélation
Aptitude de certains corps à complexer des cations suivant une structure dite en anneau.
de Chelatbildungsvermögen
Fähigkeit gewisser Körper, Kationen in einem ringförmigen Komplex zu binden.
el poder de quelación
Capacidad de ciertos cuerpos para formar complejos con cationes, según una estructura en anillo.
it potere chelante
Grado dell'attitudine di certe sostanze a complessare i cationi in una struttura ad anello.
ne metaalbindend vermogen
Het vermogen van bepaalde lichamen kationen te complexeren door het vermogen van een zgn. ringstruktuur.
pl zdolność chelatowania
Zdolność pewnych substancji do wiązania kationów z utworzeniem kompleksów pierścieniowych.

104 chelation
The formation of complexes in which the suppressed metal ion is held in a so-called ring structure, with one or more molecules having, sometimes, several electron-donor groups.

fr chélation
Formation de complexes où l'ion métal dissimulé est engagé en combinaison dite en anneau, avec une ou plusieurs molécules comportant parfois plusieurs groupements donneurs d'électrons.

de Chelatbildung
Komplexbildung, in welcher das Metall mit einem oder mehreren Molekülen, die meistens mehrere Elektronen abgebende Gruppen besitzen, ringförmig komplex gebunden ist.

el quelación
Formación de complejos, en los que el ion metálico entra en una combinación en anillo con una o varias moléculas, que a veces poseen varios grupos donantes de electrones.

it chelazione
Formazione di complessi dove l'ione metallico dissimula una struttura ad anello con una o più molecole aventi gruppi donatori di elettroni.

ne chelaatvorming
Vorming van complexen waarin het gebonden metaal-ion aanwezig is in een zogenaamde ringstruktuur met één of meer moleculen die soms meerdere electronengevende groepen bezitten.

pl chelatowanie
Tworzenie kompleksów o budowie pierścieniowej, w których metal związany jest z jedną lub więcej cząsteczkami posiadającymi niekiedy szereg grup elektrodonorowych.

105 chemical bleaching agent
A product which, by chemical action, usually oxidizing or reducing, acting under controlled conditions on textile or other materials, converts substances which affect adversely the white appearance of the material, into substances of less intense coloration.

fr agent de blanchiment chimique
Produit à action chimique, généralement oxydante ou réductrice, qui, mis en œuvre dans des conditions contrôlées sur des matières textiles ou autres, transforme en dérivés de coloration moins intense les substances ayant une influence défavorable sur l'aspect blanc.

de chemisches Bleichmittel
Produkt, welches durch seine unter kontrollierten Bedingungen erfolgende chemische (im allgemeinen oxydierende oder reduzierende) Einwirkung auf Textilfasern oder andere Materialien deren das weisse Aussehen beeinträchtigende Begleitsubstanzen in farbschwächere Verbindungen überführt.

el agente de blanqueo químico
Producto de acción química, generalmente oxidante o reductora, que actuando en condiciones controladas sobre materias textiles u otras, transforma en derivados de coloración menos intensa a las substancias que tienen una influencia desfavorable sobre su aspecto blanco.

it sbiancante chimico, candeggiante chimico
Sostanza ossidante o riducente che reagendo in condizioni controllate su tessili od altri materiali transforma in sostanze di colore meno intenso quelle che influiscono sfavorevolmente sull'aspetto bianco.

ne chemisch bleekmiddel
Een produkt dat door een chemische reactie, gewoonlijk oxydatie of reductie, onder beheerste omstandigheden inwerkt op textiel of andere materialen en stoffen die aan het witte uiterlijk van het materiaal afbreuk doen omzet in stoffen van minder intense kleur.

pl środek wybielający chemicznie
Produkt, który w wyniku prowadzonego w kontrolowanych warunkach chemicznego działania (na ogół utleniania lub redukcji) na wyroby włókiennicze lub inne produkty przeprowadza substancje wpływające niekorzystnie na biały wygląd wyrobu w inne substancje o mniej intensywnym zabarwieniu.

106 chlorine bleach
fr blanchiment en chlore
de Chlorbleiche
el blanqueo por cloro
it sbianca con cloro
ne chloorbleek
pl bielenie chlorem, bielenie chlorowe

107 chlorophyll toothpaste
fr pâte dentifrice à la chlorophylle
de Chlorophyllzahnpaste
el pasta dentífrica a la clorofila
it pasta dentifricia alla clorofilla
ne chlorofyl-tandpasta
pl chlorofilowa pasta do zębów

108 cleaning effect
fr effet nettoyant
de Reinigungswirkung
el efecto limpiador
it risultato di lavaggio
ne reinigende werking
pl działanie czyszczące

109 cleaning efficiency of toothpastes on living teeth ("in vivo")
fr pouvoir nettoyant des pâtes dentifrices sur les dents vivantes
de Reinigungseffekt von Zahnpasten auf lebende Zähne
el eficacia limpiadora de las pastas dentífricas sobre los dientes vivos
it potere detergente delle paste dentifricie sui denti vivi
ne reinigend effekt van tandpasta's *(in vivo)*
pl zdolność czyszcząca past do zębów w stosunku do żywych zębów *(in vivo)*

110 cleanser
fr nettoyant
de Reinigungsmittel (fast im Sinne von Heilmittel)
el limpiador
it detergente
ne reiniger
pl środek oczyszczający

111 cleansing cream
fr crème de nettoyage, crème de démaquillage
de Reinigungscreme
el crema limpiadora
it crema detergente
ne reinigingscrème
pl krem oczyszczający

112 cleansing effectiveness
fr efficacité de nettoyage
de Reinigungswirkung
el eficacia limpiadora
it efficacia della pulizia
ne reinigend effekt
pl skuteczność oczyszczania

113 cleansing power
fr pouvoir nettoyant
de Reinigungskraft, Reinigungsvermögen
el poder limpiador
it potere detergente
ne reinigend vermogen
pl zdolność oczyszczania

114 cloud temperature
In the case of certain non-ionic surface active agents: the temperature above which their aqueous solution becomes heterogeneous with the formation of two liquid phases.
fr température de trouble
Dans le cas de certains agents de surface non-ioniques: température au-dessus de laquelle leurs solutions aqueuses deviennent hétérogènes par formation de deux phases liquides.
de Trübungstemperatur
Im Falle gewisser nicht-ionischer Tenside: Temperatur oberhalb welcher ihre wässrige Lösung wird durch die Bildung zweier flüssiger Phasen.
el temperatura de enturbiamiento
En el caso de algunos agentes de superficie no iónicos: temperatura por encima de la cual su disolución se hace heterogénea por la formación de dos fases líquidas.
it temperatura di intorbidamento
Nel caso di alcuni tensioattivi non-ionici: temperatura al di sopra della quale le loro soluzioni acquose diventano eterogenee per formazione di due fasi liquide.
ne troebelpunt
In het geval van bepaalde niet-iogene oppervlakaktieve stoffen: bij opwarmen de temperatuur waarbij hun waterige oplossingen heterogeen worden onder de vorming van twee vloeibare fasen.
pl punkt zmętnienia
W przypadku niektórych niejonowych związków powierzchniowo czynnych temperatura powyżej której ich wodne roztwory stają się heterogeniczne w wyniku utworzenia się dwóch faz ciekłych.

115 coacervate, coacervated phase
A concentrated phase of a system which has undergone coacervation.
fr coacervat, phase coacervée
Phase concentrée d'un système ayant subi la coacervation.
de Koazervat, koazervierte Phase
Konzentrierte Phase eines der Koazervation unterworfenen Systems.

el coacervado, fase coacervada
Fase concentrada de un sistema que ha
sufrido una coacervación.

it coacervato

ne coacervaat
Geconcentreerde fase van een systeem
dat aan coacervatie onderworpen was.

pl koacerwat, faza skoacerwowana
Stężona faza układu, który uległ koacer-
wacji.

116 coacervate system
The whole of the phases of a system
which has undergone coacervation.

fr système coacervé
Ensemble des phases d'un système ayant
subi la coacervation.

de koazerviertes System
Gesamtheit der Phasen eines der Koazervation unterworfenen Systems.

el sistema coacervado
Conjunto de fases de un sistema que ha
sufrido una coacervación.

it sistema coacervato

ne gecoacerveerd systeem

pl układ skoacerwowany
Całość faz układu, który uległ koacer-
wacji.

117 coacervation
The separation into colloidal phases containing in equilibrium but in different
proportions, the same constituents.

fr coacervation
Démixtion en phases colloïdales contenant en équilibre, mais en proportions
différentes, les mêmes constituants.

de Koazervation
Entmischung einer Lösung in zwei mit-
einander im Gleichgewicht befindliche
Phasen, welche die gleichen Komponenten, jedoch in verschiedenen Mengen-
verhältnissen enthalten.

el coacervación
Separación en fases coloidales que contienen en equilibrio los mismos constituyentes, pero en proporciones diferentes.

it coacervazione

ne coacervatie
Ontmenging in colloidale fasen die met
elkaar in evenwicht zijn, maar die dezelfde komponenten in verschillende
mengverhoudingen bevatten.

pl koacerwacja
Rozdział roztworu na fazy koloidalne

zawierające te same składniki w stanie
równowagi lecz w różnych proporcjach.

118 coalescence
Joining together of droplets of an emulsion.
Notes: 1. In most cases, coalescence proceeds
through the formation of aggregates.
2. Coalescence leads to the breakdown of an
emulsion.

fr coalescence
Fusionnement de gouttelettes d'une
émulsion.
Nota: 1. Dans la plupart des cas, la coalescence
passe par la formation d'agrégats.
2. La coalescence conduit à la rupture d'une
émulsion.

de Koaleszenz
Zusammenfliessen der Tröpfchen einer
Emulsion.
Anmerkungen: 1. In den meisten Fällen ist der
Koaleszenz eine Aggregation vorgelagert.
2. Die Koaleszenz führt zum Brechen einer
Emulsion.

el coalescencia
Reunión de las gotas de una emulsión.
Observaciones: 1. En la mayoría de los casos, la
coalescencia pasa por un estado de formación de
agregados.
2. La coalescencia conduce a una ruptura de la
emulsión.

it coalescenza

ne coalescentie
Samenvloeien van de druppeltjes van
een emulsie.
Opmerkingen: 1. In de meeste gevallen gaat aan
coalescentie een aggregatie vooraf.
2. Coalescentie heeft het breken van de emulsie
tot gevolg.

pl koalescencja
Łączenie się cząstek (kropelek) emulsji.
Uwagi: 1. W większości przypadków koalescencja zachodzi poprzez etap tworzenia się agre-
gatów.
2. Koalescencja prowadzi do rozbicia emulsji.

119 cold process soap
Soap obtained by the saponification of
molten fatty matter mixed with a concentrated alkali lye in the cold. The heat
produced by the exothermic reaction
leads to completion of the reaction.

fr savon d'empâtage à froid
Savon obtenu par saponification de corps
gras fondus, mélangés avec une lessive
alcaline concentrée et froide. La chaleur
produite par la réaction exothermique
mène au bout la réaction.

de kaltgerührte Seife
Seife, die durch Verseifung der geschmol-
zenen Fette (meist Kokos- oder Palm-
kernöl) beim Vermischen mit kalter,

konzentrierter Alkalilauge hergestellt wird. Die bei der exothermen Reaktion freiwerdende Wärme dient zur Vervollständigung der Verseifung.

el jabón de empaste en frío
Jabón obtenido par saponificación de grasas fundidas (generalmente, aceites de coco y de palmiste) mezcladas con una lejía alcalina concentrada y fría. El calor producido por la reacción exotérmica lleva a su término la reacción.

it sapone d'impasto a freddo
Sapone ottenuto a mezzo saponificazione di prodotti grassi fusi, miscelati con lisciva di alcali caustico concentrato a freddo. Il calore prodotto dalla reazione esotermica porta al completamento della reazione.

ne koudgeroerde zeep
Zeep, die door verzeping van gesmolten vetten (meest kokos- of palmkernolie) bij het mengen met koude, geconcentreerde loog bereid wordt. De bij de exotherme reactie vrijkomende warmte dient ter vervolmaking van de verzeping.

pl mydło wykonane przez zmydlanie na zimno
Mydło otrzymane przez zmydlenie stopionego tłuszczu (najczęściej oleju kokosowego lub z ziarn palmowych) w wyniku zmieszania go z zimnym, stężonym ługiem. Ciepło wytworzone w wyniku tej egzotermicznej reakcji zapewnia doprowadzenie procesu zmydlania do końca.

120 cold wave
fr ondulation obtenue à froid, permanente à froid
de Kaltwelle
el ondulación en frío
it ondulazione a freddo
ne haarkrullen bij gewone temperatuur
pl ondulacja na zimno

121 cold waving of hair
fr permanente à froid
de Kaltwellen des Haares
el permanente en frío
it permanente a freddo
ne haargolven bij gewone temperatuur
pl ondulowanie włosów na zimno

122 collapsible tube
fr tube souple

de Falttube
el tubo flexible, tubo plegable
it tubetto dentifricio
ne samendrukbare tube
pl tuba (do past lub kremów)

123 coloration
fr coloration
de Färbung
el coloración
it colorazione
ne kleuring
pl farbowanie, koloryzowanie

124 colour shampoo
fr shampooing colorant
de Haartönung
el champu colorante
it shampoo colorante
ne kleurshampoo
pl szampon koloryzujący

125 coloured rinse, tint rinse
fr rinçage colorant
de Tönungsspülung
el lavado colorante
it lavaggio colorante, reflex
ne kleurend spoelmiddel
pl płukanka koloryzująca

126 coloured wash
fr lavage du linge de couleur, lavage des articles colorés
de Buntwäsche
el lavado de prendas de color
it lavaggio di tessuti colorati
ne bontwas
pl bielizna kolorowa, kolory (do prania)

127 compatibility of the skin
fr compatibilité de la peau, tolérance de la peau
de Hautverträglichkeit
el compatibilidad de la piel, tolerancia de la piel
it compatibilità con la pelle, tolleranza della pelle
ne verdraagbaarheid van de huid, verenigbaarheid van de huid
pl tolerancja skóry

128 complexing efficiency
fr pouvoir complexant
de Fähigkeit zur Komplexbildung, Komplexbildungsvermögen
el poder de formación de complejos, poder secuestrante
it potere chelante

ne vermogen tot complexvorming
pl skuteczność kompleksowania

129 complexing power
The ability of certain bodies to combine with cations which then lose their ionic character.
fr pouvoir complexant
Aptitude de certains corps à fixer des cations, qui peuvent perdre ainsi leur identité ionique.
de Komplexbildungsvermögen
Fähigkeit gewisser Körper, Kationen so zu binden, dass sie ihre chemischen Eigenschaften verändern können.
el poder complexante
Capacidad de ciertos cuerpos para fijar cationes de tal forma que puedan perder su identidad iónica.
it potere complessante
Grado dell'attitudine di certe sostanze a fissare cationi che perdono conseguentemente la loro forma ionica.
ne complexvormend vermogen
Het vermogen van bepaalde lichamen zich te verbinden met kationen die dan hun ion-karakter verliezen.
pl zdolność kompleksowania
Zdolność pewnych substancji do wiązania kationów, w wyniku czego tracą one charakter jonowy.

130 complexion
fr teint
de Teint, Gesichtsfarbe
el tez
it carnagione
ne gelaatskleur
pl cera, karnacja

131 complexion
The transforming of a metal cation into a new ion by the action of molecules containing at least one electron-donor group.
fr complexion
Transformation d'un cation métallique en un ion nouveau, par l'intervention de molécules comportant au moins un groupement donneur d'électrons.
de Komplexbildung
Bildung eines neuen Ions aus einem Metallkation durch Einwirkung von Molekülen, die mindestens eine Elektronen abgebende Gruppe besitzen.
el complexión
Transformación de un catión metálico

en un ion nuevo, por la intervención de moléculas que contienen por lo menos un grupo donante de electrones.
it complessione
Trasformazione di un catione metallico in un nuovo ione per azione di molecole con almeno un gruppo donatore di elettroni.
ne complexvorming
De omzetting van een kation in een nieuw metaalion door de werking van moleculen met tenminste één electron-donorgroep.
pl kompleksowanie
Przekształcenie kationu metalu w nowy jon w wyniku oddziaływania cząsteczek zawierających co najmniej jedną grupę elektrodonorową.

132 concurrent process
fr procédé en courants parallèles, procédé dans le courant
de Gleichstromverfahren
el proceso a concorriente
it processo in equicorrente
ne werkwijze met gelijkstroom
pl proces współprądowy

133 condensed phosphates
fr phosphates condensés
de kondensierte Phosphate
el fosfatos condensados
it fosfati condensati
ne gecondenseerde fosfaten
pl zespolone fosforany

134 contact angle
In a plane perpendicular to the line of separation formed by three phases, the tangents to the two curves formed when the plane cuts one of the three phases are considered. The angle formed by these tangents is called the contact angle of this phase in relation to the other two.
fr angle de contact
Dans un plan perpendiculaire à la ligne de séparation formée par trois phases, on considère les tangentes aux deux courbes suivant lesquelles le plan coupe l'une des trois phases. L'angle formé par ces tangentes est appelé angle de contact de cette phase par rapport aux deux autres.
de Kontaktwinkel
In einer Ebene, welche auf der durch

drei Phasen gebildeten Trennlinie senkrecht steht, betrachtet man die an die beiden eine Phase begrenzenden Kurven gelegten Tangenten. Der durch diese Tangenten gebildete Winkel heisst Kontaktwinkel dieser Phase in Bezug auf die zwei anderen.

el ángulo de contacto
En un plano perpendicular a la línea de separación de tres fases, se consideran las tangentes a las dos curvas según las cuales el plano corta a una de las tres fases. El ángulo formado por estas tangentes se llama ánguio de contacto de dicha fase con respecto a las otras dos.

it angolo di contatto
In un piano perpendicolare alla linea di separazione formata da 3 fasi si considerino le tangenti alle due curve secondo le quali il piano taglia una delle 3 fasi. L'angolo formato da queste tangenti è chiamato angolo di contatto in relazione alla fase delle altre due.

ne randhoek
In een vlak dat loodrecht staat op de scheidingslijn gevormd door drie fasen gaat het om de raaklijnen aan de twee krommen die ontstaan wanneer het vlak één van de drie fasen snijdt. De hoek gevormd door deze raaklijnen wordt de randhoek van deze fasen ten opzichte van de twee andere genoemd.

pl kąt zwilżania
Jeśli na płaszczyźnie prostopadłej do linii podziału utworzonej przez trzy fazy rozpatrywane są styczne do krzywych rozdzielających dwie fazy, to kąt utworzony przez te styczne zwie się kątem zwilżania danej fazy w stosunku do dwóch pozostałych.

135 cooling press
fr presse réfrigérante
de Kühlpresse
el prensa de enfriamiento
it pressa refrigerante
ne koelpers
pl chłodnia płytowa

136 cooling roller mill
fr appareil réfrigérant à cylindres
de Kühlwalzwerk
el cilindros de enfriamiento, cilindros refrigerantes

it cilindri raffreddatori
ne koelwals
pl walce chłodzące do mydła

137 corium, dermis
fr derme
de Corium
el dermis
it derma
ne lederhuid
pl skóra właściwa

138 cosmetic cream
fr crème cosmétique, crème de beauté
de kosmetische Creme
el crema cosmética, crema de belleza
it crema di bellezza
ne cosmetische crème
pl krem kosmetyczny

139 countercurrent process
fr procédé à contrecourant
de Gegenstromverfahren
el proceso a contracorriente
it processo in controcorrente
ne werkwijze in tegenstroom
pl proces przeciwprądowy

140 covering power
fr pouvoir couvrant
de Deckkraft
el poder cubriente
it potere coprente
ne dekkracht
pl zdolność krycia

141 cracking of soap
fr fendillement du savon, craquelage
de Rissigwerden der Seifen, Rissbildung der Seifen
el resquebrajamiento, rajadura, cuarteamiento
it screpolatura del sapone
ne barsten van de zeep
pl pękanie mydła

142 creamy liquid powder base
fr base de poudre (crème liquide)
de halbflüssige Pudergrundlage
el base de polvos (crema líquida)
it crema liquida (a base di cipria)
ne poederbasis (vloeibare crème)
pl półpłynna baza pudrowa, półpłynna podstawa pudrowa

143 crease resistant, uncreasable
fr infroissable
de knitterfrei

el inarrugable
it ingualcibile
ne kreukvrij
pl niemnący, niegniotący

144 creasing resistance, crease resistance
fr infroissabilité
de Knitterfestigkeit, Falzfestigkeit
el inarrugabilidad
it ingualcibilità
ne kreukvrij zijn
pl odporność na gniecenie

145 critical concentration for micelle formation (symbol c_M)
Characteristic concentration of surface active agents in solution (in practice a narrow range of concentration) above which the appearance and development of micelles bring about a sudden variation in the relation between the concentration and certain physico-chemical properties. The critical concentration for micelle formation is determined by the point of intersection of two extrapolated curves which represent this relation above and below the critical concentration.

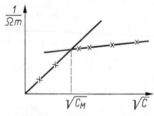

Note: The value of the critical concentration for micelle formation can depend to a certain extent on the property under consideration and on the method chosen to measure this property. The graph explains this definition, which represents the variations of a physico-chemical property (electrical conductivity) dependent on the concentration.

fr concentration critique pour la formation de micelles (symbole c_M)
Concentration caractéristique des agents de surface en solution (pratiquement étroit intervalle de concentration) au-dessus de laquelle l'apparition et le développement de micelles provoquent une variation brusque dans la relation entre la concentration et certaines propriétés physico-chimiques. La concentration critique pour la formation de micelles est déterminée par le point d'intersection des deux courbes extra-polées qui représentent cette relation au-dessus et au-dessous de la concentration critique.

Nota: La valeur de la concentration critique pour la formation de micelles peut dépendre dans une certaine mesure de la propriété considérée et de la méthode choisie pour mesurer cette propriété. Le graphique, indiquant la variation d'une propriété physico-chimique (conductivité électrique) en fonction de la concentration, explicite cette définition.

de kritische Micellbildungskonzentration (Symbol c_M)
Charakteristische Konzentration (praktisch: enger Konzentrationsbereich) von Tensiden in Lösung, oberhalb welcher die Zahl der Micellen in solchem Masse ansteigt, dass hierdurch die Konzentrationsabhängigkeit gewisser physikalischer Eigenschaften sprunghaft verändert wird. Die kritische Micellbildungskonzentration ist festgelegt durch den Schnittpunkt zweier extrapolierter Kurvenzüge, welche die Eigenschaft der Lösung als Funktion der Konzentration beschreiben.

Anmerkung: Der Wert der kritischen Micellbildungskonzentration kann bis zu einem gewissen Grade von der Eigenschaft abhängen, die bei der Ermittlung zugrunde gelegt wurde. Die Abbildung zeigt die Änderung einer physikalischen Eigenschaft (hier elektrische Leitfähigkeit) als Funktion der Tensidkonzentration.

el concentración crítica de formación de micelas
Concentración característica de los agentes de superficie en disolución (prácticamente, estrecho intervalo de concentración), por encima de la cual la aparición y desarrollo de micelas provoca una variación brusca de la relación entre la concentración y ciertas propiedades físico-químicas. La concentración crítica de formación de micelas viene determinada por el punto de intersección de las curvas extrapoladas, que representan esta relación, por encima y por debajo de la concentración crítica.

Observación: El valor de la concentración crítica de formación de micelas puede depender en cierta medida de la propiedad considerada y del método elegido para medir esta propiedad.

it concentrazione critica micellare
Concentrazione caratteristica dei tensioattivi in soluzione al di sopra della quale una parte delle molecole e/o ioni dei tensioattivi si raccoglie in micelle disperse nella soluzione.

Nota: Tale concentrazione corrisponde in pratica alla comparsa di una quantità di micelle sufficiente a modificare in modo sensibile certe proprietà chimico-fisiche della soluzione. In caso di miscele la valutazione della concentrazione critica micellare può dipendere dal metodo di determinazione usato.

ne kritische micel concentratie
Karakteristieke concentratie van oppervlakaktieve stoffen in oplossing (in de praktijk een beperkt concentratiegebied), boven welke het ontstaan en zich ontwikkelen van micellen een plotselinge wijziging in het verband tussen de concentratie en bepaalde fysisch-chemische eigenschappen teweegbrengt. De kritische concentratie door micelvorming wordt bepaald voor het snijpunt van twee geëxtrapoleerde curven, die dit verband boven en beneden de kritische concentratie weergeven.

Opmerking: De waarde van de kritische concentratie voor micelvorming kan tot op zekere hoogte afhankelijk zijn van de beschouwde eigenschap en van de voor de meting van deze eigenschap gekozen methode. De grafiek, die de variatie aangeeft van een fysisch-chemische eigenschap (electrische geleidbaarheid) als functie van de concentratie, verklaart deze definitie.

pl krytyczne stężenie powstawania miceli (symbol c_M)
Charakterystyczne stężenie związków powierzchniowo czynnych w roztworze (w praktyce wąski zakres stężeń), powyżej którego pojawienie się i wzrost liczby miceli powoduje nagłe zmiany zależności pomiędzy stężeniem a pewnymi własnościami fizykochemicznymi. Krytyczne stężenie powstawania miceli określane jest przez punkt przecięcia się dwóch ekstrapolowanych krzywych, które opisują tę zależność poniżej i powyżej stężenia krytycznego.

Uwaga: Wartość krytycznego stężenia powstawania miceli może zależeć w pewnym stopniu od rodzaju rozpatrywanej własności i od metody wybranej do jej pomiaru. Wykres objaśniający tę definicję przedstawia zmiany własności fizycznej (przewodnictwa elektrycznego) w zależności od stężenia związku powierzchniowo czynnego.

146 critical micelle concentration (c.m.c.)
fr concentration critique de micelles, concentration critique micellaire, concentration critique de formation de micelles
de kritische Micellkonzentration
el punto crítico de la formación de micelas, concentración crítica micelar
it concentrazione critica delle micelle, concentrazione critica di formazione delle micelle
ne kritische micelconcentratie (k.m.c.), kritische micelvormingsconcentratie
pl krytyczne stężenie micelarne

147 cross-linking agent
fr agent produisant des liaisons transversales
de vernetzende Verbindung
el agente productor de uniones transversales
it agente per la produzione dei legami trasversali
ne middel voor dwarsverbindingen
pl czynnik sieciujący

148 crow's feet
fr patte d'oie
de Krähenfüsse
el patas de gallo
it zampe d'oca
ne kraaiepootjes
pl kurze łapki

149 curd soap, soap on lye†
The phase of soap in the pan, appearing in the form of flocks or grains of concentrated soap made up of opaque crystalline fibres in equilibrium with a practically soap free lye.
fr savon grainé, savon levé sur lessive
Phase du savon en chaudière, se présentant sous forme de flocons ou "grains" de savon concentré, constitués par des fibres cristallines opaques, se trouvant en équilibre avec une lessive ne contenant pratiquement pas de savon.
de geronnener Seifenkern, Seifenkern auf Unterlauge
Seifenphase in Form konzentrierter Seifenflocken oder "Kerne", die aus opaken krystallinen Fasern bestehen und mit einer praktisch seifenfreien Lauge (Unterlauge) im Gleichgewicht stehen.
el jabón graneado
Fase anisótropa y opaca del jabón líquido concentrado, en equilibrio con una lejía que prácticamente no contiene jabón.
it sapone granito, sapone levato su liscivia
Fase del sapone in caldaia, sotto forma di precipitato flocculoso o in grumi di sapone concentrato, costituito da parti cristalline opache che si trovano in equilibrio con una liscivia praticamente esente da sapone.
ne uitgezouten zeep, kernzeep op onderloog

Zeepfase in de vorm van geconcentreerde zeepvlokken of kernen, die uit opake kristallijnen vezels bestaan en met een praktisch zeepvrije loog (onderloog) in evenwicht staat.

pl mydło rdzeniowe
Faza mydła występująca w kotle warzelnym, pojawiająca się w formie kłaczków lub ziarna stężonego mydła, tworząca nieprzeźroczyste, krystaliczne włókienka, znajdujące się w równowadze z praktycznie wolnym od mydła ługiem spodnim.
† See appendix

150 curler
fr bigoudi
de Lockenwickler
el bigudí
it bigodini
ne krulpen
pl lokówka

151 cuticle
fr cuticule
de Cuticula
el cutícula
it cuticola
ne nagelriem
pl naskórek

152 cutter
fr coupeuse
de Schneidemaschine
el cortadora
it taglierina
ne snijmachine
pl krajarka

153 cutting oil
A lubricating compound, emulsifiable or not, which facilitates the work of machine tools and the dissipation of heat produced.
Note: It may contain an additive imparting anti-corrosion properties.

fr huile de coupe
Composition lubrifiante, émulsionnable ou non, facilitant le travail des outils de coupe des métaux et l'évacuation de la chaleur produite par ce travail.
Nota: Elle peut renfermer un additif lui communiquant un pouvoir d'anticorrosion.

de Schneidöl
Emulgierbare oder nicht emulgierbare Mischung, welche die Arbeit der Schneidwerkzeuge sowie die Ableitung der hierbei erzeugten Wärme erleichtert.
Anmerkung: Schneidöle können Zusatzstoffe enthalten, welche ihnen Antikorrosionsvermögen verleihen.

el aceite de corte
Composición lubrificante, emulsionable o no, que facilita el trabajo de las herramientas de corte de metales y la evacuación del calor producido por este trabajo.
Observación: Puede contener un aditivo que le comunique un poder de anticorrosión.

it —
ne snij-olie
Een smeermiddel, al of niet emulgeerbaar, dat het werken met machinale werktuigen vergemakkelijkt en de gevormde warmte wegneemt.

pl płyn obróbkowy (koncentrat)
Substancja smarna, emulgująca lub nieemulgująca, ułatwiająca pracę narzędzia skrawającego oraz ułatwiająca odprowadzenie ciepła wytworzonego przy obróbce skrawaniem.
Uwaga: Płyn ten może zawierać dodatki antykorozyjne.

154 cystine
fr cystine
de Cystin
el cistina
it cistina
ne cystine
pl cystyna

D

155 **damage to fibre, weakening of the fibre**
fr attaque de la fibre, usure de la fibre, dégat dans la fibre, détérioration des fibres, affaiblissement de la fibre
de Faserschäden, Faserschädigung, Faserschwächung, Faserabbau
el ataque de la fibra, dañado de la fibra, deterioración de la fibra, acción deteriodora para la fibra
it attaco della fibra, usura della fibra, deteriorazione della fibra, indebolimento della fibra
ne aantasting van de vezel
pl uszkodzenie włókna, osłabienie włókna

156 **dandruff formation**
fr formation de pellicule
de Schuppenbildung
el formación de caspa
it formazione di forfora
ne roosvorming
pl tworzenie się łupieżu

157 **day cream**
fr crème de jour
de Tagescreme
el crema de día
it crema per giorno
ne dagcrème
pl krem na dzień

158 **dead tissue remover**
fr produit pour enlever les tissus morts
de Hornhautentferner
el producto para quitar los tejidos muertos
it prodotto per rimuovere i tessuti morti
ne middel tot verwijdering van het dode weefsel
pl środek do złuszczania warstwy rogowej naskórka

159 **decomposition**
fr décomposition
de Zersetzung
el decomposición
it decomposizione
ne ontleding
pl rozkład, rozpad

160 **defatting of the skin**
fr dégraissage de la peau
de Hautentfettung
el desengrasado de la piel
it sgrassaggio della pelle

ne ontvetting van de huid
pl odtłuszczanie skóry

161 **deforming**
fr déformation
de Verformen
el deformación
it deformazione
ne vervorming
pl odkształcanie, deformacja

162 **degreasing power**
fr pouvoir dégraissant
de Entfettungsvermögen
el poder desengrasante
it potere sgrassante
ne ontvettend vermogen
pl zdolność odtłuszczająca

163 **degree of biodegradation**
The extent of the loss of surface activity determined after biodegradation treatment of a surface active agent in solution.

Notes: 1. The degree of biodegradation depends, within acceptable limits, on the relevant test method.
2. The degree of biodegradation is usually evaluated by representative analytical determinations.

fr degré de biodégradation
Taux de perte d'activité de surface constatée après une opération de biodégradation d'un agent de surface en solution.

Nota: 1. Le degré de biodégradation dépend, dans des limites acceptables, de la méthode d'essai appliquée.
2. Le degré de biodégradation est évalué généralement par des déterminations analytiques représentatives.

de biologischer Abbaugrad
Auf den Ausgangswert bezogene Verminderung der Konzentration einer dem biologischen Abbau unterworfenen Tensidlösung.

Anmerkung: Der biologische Abbaugrad wird im allgemeinen nach den massgebenden Bestimmungsmethoden für Tenside ermittelt.

el grado de biodegradación
Porcentaje de pérdida de actividad de superficie observada después de una operación de biodegradación de un agente de superficie en disolución.

Observaciones: 1. El grado de biodegradación depende, dentro de límites aceptables, del método de ensayo aplicado.
2. El grado de biodegradación se evalúa generalmente por medio de determinaciones analíticas representativas.

it grado di biodegradazione
Tasso di perdita di tensioattività determinato dopo trattamento di biodegradazione di tensioattivo in soluzione acquosa.
Nota: 1. Il grado di biodegradazione dipende, entro limiti accettabili, dal metodo di prova usato.
2. Il grado di biodegradazione è normalmente misurato per mezzo di determinazioni analitiche rappresentative.

ne biodegradatiegraad
Waargenomen verlies aan oppervlakaktiviteit nadat een oppervlakaktieve stof in oplossing onderworpen is geweest aan omstandigheden die biodegradatie kunnen veroorzaken.
Opmerkingen: 1. De biodegradatiegraad is binnen bepaalde grenzen afhankelijk van de gebruikte bepalingsmethode.
2. De biodegradatiegraad wordt in het algemeen bepaald door geschikte analytische methoden.

pl stopień rozkładu biologicznego, stopień biodegradacji
Zakres utraty własności powierzchniowo czynnych oznaczany po rozkładzie biologicznym związku powierzchniowo czynnego
Uwaga: 1. Stopień rozkładu biologicznego zależy, w dopuszczalnych granicach, od stosowanej metody badania.
2. Stopień rozkładu biologicznego zazwyczaj oceniany jest poprzez odpowiednie oznaczenia analityczne.

164 degree of soiling
fr degré de salissure, degré de souillure
de Verschmutzungsgrad
el grado de suciedad
it grado di sporco
ne graad van bevuiling
pl stopień zabrudzenia

165 dehydration
(*a*) Physical operation resulting in the removal of all or part of the water bound to a product.
(*b*) Chemical reaction resulting in the removal of one or more molecules of water.
fr déshydratation
(*a*) Opération physique permettant d'éliminer tout ou partie de l'eau liée à un produit.
(*b*) Réaction chimique permettant d'éliminer une ou plusieurs molécules d'eau.
de Dehydratation
(*a*) Physikalische Operation, bei welcher das an ein Produkt gebundene Wasser teilweise oder vollständig entfernt wird.

(*b*) Chemische Reaktion, bei der ein oder mehrere Mol Wasser abgespalten werden.
el deshidratación
(*a*) Operación física, en la que se elimina, total o parcialmente, el agua contenida en un producto.
(*b*) Reacción química en la que se eliminan una o varias moléculas de agua.
it disidratazione
(*a*) Operazione fisica che consente di eliminare tutta o in parte l'acqua legata ad un prodotto.
(*b*) Reazione chimica che consente di eliminare una o più molecole d'acqua.
ne dehydratering
(*a*) Fysische werkwijze die leidt tot eliminering van al het aan een produkt gebonden water of een deel ervan.
(*b*) Chemische reactie die leidt tot eliminering van een of meer moleculen kristalwater.
pl dehydratacja, odwodnienie
(*a*) Proces fizyczny, w wyniku którego następuje częściowe lub całkowite usunięcie wody związanej w produkcie.
(*b*) Reakcja chemiczna, w wyniku której następuje odszczepienie jednej lub więcej cząsteczek wody.

166 delta phase
The crystalline form which tends to be formed by the cooling of high molecular weight soaps with high water content (over 50%) or by exerting a mechanical force on soaps of this nature in the beta phase at temperatures between 10 and 15°C. The delta phase has a solution rate intermediate between that of the beta and omega phases, but it is less firm than these two phases.
fr phase delta
Forme cristalline, qui tend à prendre naissance par refroidissement des savons à poids moléculaire élevé et haute teneur en eau (plus de 50%), ou en exerçant à des températures entre 10 et 15°C une contrainte mécanique sur des savons de ce genre en phase bêta. La phase delta a une vitesse de dissolution intermédiaire entre celle des phases bêta et oméga, mais est moins ferme que ces deux phases.
de Delta-Phase
Kristalline Form, die durch Abkühlung

von Seifen höheren Molekulargewichts und hohen Wassergehalts (mehr als 50%), oder durch mechanische Bearbeitung solcher in Form der Beta-Phase vorliegenden Seifen, bei Temperaturen zwischen 10 und 15°C entsteht. Die Lösungsgeschwindigkeit der Delta-Phase liegt zwischen der der Beta- und Omega-Phase; die Delta-Phase ist weicher als diese beiden Phasen.

el fase delta
Forma cristalina que se produce por enfriamiento de jabones de peso molecular elevado y de alto contenido de agua (más de 50%) o por la acción de un esfuerzo mecánico, a temperaturas entre 10 y 15°C, sobre jabones de este género en fase beta. La velocidad de disolución de la fase delta es intermedia entre la de las fases beta y omega, pero la fase delta es más blanda que estas dos.

it fase delta
La forma cristallina, che si ottiene con il raffreddamento di saponi ad alto peso molecolare e ad alto contenuto di acqua (oltre il 50%) o con azione meccanica su saponi di questa specie nella fase beta a temperature fra i 10 e i 15°C. La fase delta ha una solubilità intermedia fra le fasi beta e omega, ma è meno compatta di queste due fasi.

ne delta-fase
De kristallijnen vorm, die door afkoeling van zepen met hoog moleculair-gewicht en hoog watergehalte (meer dan 50%), of door mechanische bewerking van zulke in de beta-fase verkerende zepen, bij temperaturen tussen 10 en 15°C ontstaat.

pl faza delta
Krystaliczna postać mydła, mająca tendencję do powstawania bądź w wyniku chłodzenia wielkocząsteczkowych mydeł o dużej zawartości wody (ponad 50%), bądź w wyniku mechanicznej obróbki mydeł tego typu występujących w fazie beta w zakresie temperatur 10–15°C. Szybkość rozpuszczania fazy delta leży pomiędzy szybkościami rozpuszczania faz beta i omega, z tym jednak, że jest ona bardziej miękka niż obie pozostałe fazy.

167 **dental caries**
fr carie des dents
de Zahnkaries
el caries dental
it carie dentaria
ne tandcariës
pl próchnica zębów

168 **dental finisher**
fr tablette dentifrice purifiante
de Zahnreinigungstablette
el tableta dental purificadora
it tavoletta dentifricia purificante
ne tandreinigend tablet
pl tabletki do czyszczenia zębów

169 **denture cleaner**
fr produit pour nettoyer les appareils dentaires
de Gebissreinigungsmittel (künstliche Gebisse)
el producto para limpiar los aparatos dentales
it prodotto per pulire gli apparecchi dentari
ne reinigingsmiddel voor kunstgebitten
pl środek do czyszczenia protez zębowych

170 **deodorant**
fr désodorisant
de desodorisierend
el desodorante
it deodorante
ne reukverdrijvend middel
pl dezodorant, środek dezodorujący

171 **deodorant soap**
Toilet soap with bacteriostatic additives, which have a deodorant effect.
fr savon désodorant
Savon de toilette contenant des additifs bactériostatiques qui ont comme conséquence une action désodorante.
de desodorierende Seife, Deodorantseife
Feinseife (Toiletteseife) mit keimhemmenden (bakteriostatischen) Zusätzen, wodurch eine desodorierende Wirkung erreicht wird.
el jabón desodorante
Jabón de tocador con aditivos bacteriostáticos, que producen una acción desodorante.
it sapone deodorante
Sapone da toeletta con additivi batteriostatici ad effetto deodorante.
ne desodoriserende zeep
Toiletzeep met een bacteriostatische

substantie, die een desodoriserend effekt heeft.
pl mydło dezodorujące
Mydło toaletowe zawierające dodatki bakteriobójcze o działaniu dezodorującym.

172 deodorant stick
fr bâton désodorisant
de desodorisierender Stift
el barrita desodorante
it "stick" deodorante
ne reukverdrijvende staaf
pl sztyft dezodorujący, dezodorant w sztyfcie

173 deodorising compound, deodorant
fr déodorisant
de Desodorierungsmittel
el desodorante
it deodorante
ne deodoriseermiddel
pl środek dezodorujący, dezodorant

174 depth of penetration
fr profondeur de pénétration
de Eindringstiefe
el profundidad de penetración
it profondità di penetrazione
ne diepte van indringing
pl głębokość penetracji, głębokość wnikania

175 dermal change
fr transformation du derme
de Hautveränderung
el cambio dérmico
it modificazione della pelle
ne verandering van de huid
pl zmiany skórne

176 dermatology
fr dermatologie
de Dermatologie
el dermatología
it dermatologia
ne leer der huidziekten
pl dermatologia

177 desizing assistant and assistant for removal of printing thickeners
Product which accelerates the removal of print thickeners as well as the desizing insofar as starch products are concerned. To this end, appropriate surface active agents such as wetting agents or detergents can be added to agents based on enzymes.
fr adjuvant de désencollage et adjuvant

pour l'élimination des épaississants d'impression
Produit qui accélère l'élimination des épaississants d'impression, ainsi que le désencollage, pour autant qu'il s'agisse de produits amylacés. Dans ce but, des agents de surface appropriés tels que des agents mouillants ou détergents peuvent être ajoutés aux préparations enzymatiques.
de Entschlichtungshilfsmittel und Hilfsmittel zur Entfernung von Druckverdickungen
Produkt, das den Entschlichtungsprozess und das Entfernen von Druckverdickungsmitteln — sofern stärkehaltige Produkte vorliegen — beschleunigt. Beim Arbeiten mit Enzympräparaten können auch geeignete grenzflächenaktive Stoffe, d.h. Netz- oder Waschmittel, mitverwendet werden.
el auxiliar de desencolado y de eliminación de espesantes de estampación
Producto que, en presencia de substancias amiláceas, acelera el desencolado y la eliminación de los espesantes de estampación. Si se utilizan preparaciones enzimáticas, pueden añadirse agentes de superficie apropiados, tales como agentes humectantes o detergentes. La eliminación del aceite de linaza, o de encolados que lo contengan, se facilita por el empleo anteriormente indicado de agentes de superficie, asociados a disolventes y/o a agentes oxidantes.
it —
ne hulpmiddel voor het verwijderen van pap en drukinktverdikkingsmiddelen.
Een produkt dat het verwijderen van pap en drukinktverdikkers bevordert.
pl wspomagacz odklejania (usuwania klejonki) i środek pomocniczy do usuwania zagęstników farb drukarskich
Produkt przyśpieszający usuwanie zagęstników farb drukarskich oraz odklejanie klejonek skrobiowych. W tym celu do preparatów enzymatycznych można dodawać odpowiednie związki powierzchniowo czynne (zwilżające i piorące).

178 detergency, detergence
The process by which soil is dislodged and brought into a state of solution or dispersion. In its usual sense, detergency has the effect of cleaning surfaces. It is

the result of the action of several physico-chemical phenomena.

fr détergence
Processus selon lequel des salissures (souillures) sont enlevées et mises en solution ou en dispersion. Au sens ordinaire, la détergence a pour effet le nettoyage des surfaces. Elle est la résultante de la mise en œuvre de plusieurs phénomènes physico-chimiques.

de Waschvorgang
Vorgang, bei dem Schmutz oder Anschmutzungen entfernt und in Lösung oder Dispersion übergeführt werden. Im allgemeinen bewirkt der Waschvorgang die Reinigung von Oberflächen. Er ist das Ergebnis der Wirkung mehrerer physikalisch-chemischer Vorgänge.

el detergencia
Proceso de eliminación de la suciedad y de su puesta en disolución o dispersión. En su sentido corriente, la detergencia tiene por objeto la limpieza de superficies y es la resultante de la acción de varios fenómenos físico-químicos.

it detergenza
Processo per mezzo del quale lo sporco viene rimosso e portato in soluzione o dispersione. Nel suo significato comune la detergenza ha l'effetto, di rendere pulite le superfici. E' la risultante di diversi fenomeni chimico-fisici concomitanti.

ne waswerking
Het proces waarbij vuil wordt losgemaakt en in een opgeloste of gedispergeerde toestand wordt gebracht. In de normale betekenis leidt de waswerking tot de reiniging van oppervlakken. Het is het resultaat van de werking van verschillende fysisch-chemische verschijnselen.

pl proces prania, proces czyszczenia
Proces, w którym brud zostaje usunięty i przeprowadzony w stan roztworu lub zawiesiny. Ogólnie biorąc działanie piorące prowadzi do oczyszczenia powierzchni. Proces ten jest wynikiem szeregu zjawisk fizyko-chemicznych.

179 detergent, washing agent†
A product specially formulated to promote the development of detergency.
Note: A detergent comprises essential components (surface active agents), and generally complementary components (builders, etc.).

fr détergent, produit de lavage
Produit dont la composition pratique est spécialement étudiée pour concourir au développement des phénomènes de détergence.
Nota: Un détergent comprend des composants essentiels (agents de surface) et, généralement, des composants complémentaires (adjuvants, etc.).

de Waschmittel
Ein Produkt, dessen für praktische Zwecke bestimmte Zusammensetzung dahingehend entwickelt worden ist, im Waschvorgang die höchste Wirkung zu erzielen.
Anmerkung: Ein Waschmittel enthält eine aktive Substanz (grenzflächenaktive Verbindung), zusätzlich gegebenenfalls ergänzende Bestandteile (Builder, Verstärker, Füllstoffe, etc.).

el detergente, producto de lavado
Producto cuya composición ha sido establecida primordialmente para el desarrollo de los fenómenos de detergencia.
Observación: Un detergente es una formulación que posee componentes esenciales (agentes de superficie) y componentes complementarios (coadyuvantes, reforzadores, cargas y aditivos).

it detergente
Prodotto particolarmente studiato nella sua formulazione pratica per promuovere la detergenza.
Nota: Un detergente comprende costituenti principali (tensioattivi) e costituenti complementari (adiuvanti, rafforzatori, sostanze di carica e additivi).

ne wasmiddel
Een produkt waarvan de samenstelling speciaal bedoeld is voor het bevorderen van de ontwikkeling van waswerking.
Opmerking: Een wasmiddel bevat oppervlakaktieve stoffen en toevoegingen.

pl środek piorący
Produkt o składzie specjalnie dobranym pod kątem uzyskania najlepszego działania piorącego.
Uwaga: W skład środka piorącego wchodzą składniki podstawowe (związki powierzchniowo czynne) oraz składniki dodatkowe (aktywne wypełniacze itd.).
† See appendix

180 detergent action, washing performance, cleaning efficiency
fr action détergente
de Waschwirkung
el acción detergente
it azione detergente
ne waswerking
pl działanie piorące

181 detergent characteristics
fr propriétés détergentes

de Wascheigenschaften
el propiedades detergentes, características detergentes
it proprietà detergenti
ne waseigenschappen
pl własności piorące

182 detergent effect
fr effet détergent
de Wascheffekt
el efecto del lavado, efecto del detergente
it effetto detergente
ne waseffekt
pl efekt wyprania

183 detergent for the coloured wash
fr détergent pour articles colorés
de Buntwaschmittel
el detergente para los tejidos colorados, detergente para ropa de color
it detergente per articoli colorati
ne wasmiddel voor de bonte was
pl środek do prania bielizny kolorowej (kolorów)

184 detergent for the textile industry
Product used in the textile industry to eliminate fats and soiling matter on textiles during manufacture and finishing. Its composition and/or its formulation meet the requirements for the various stages of the work, for example for scouring raw wool, yarns or pieces and for the back-washing of dyed and printed fabrics, etc.
Note: These products are surface active agents or compounds comprising them, such as: soaps, alkylsulphates, alkylsulphonates, fatty acid condensates, polyglycol esters and ethers.

fr agent détergent pour l'industrie textile
Produit servant, dans l'industrie textile, à éliminer les matières grasses et les souillures des textiles au cours de la fabrication et du finissage. Sa composition et/ou sa formulation doivent répondre aux exigences du stade de travail considéré, par exemple, lavage de la laine brute, des filés ou des pièces, dans le lavage subséquent des teintures et des impressions, etc.
Nota: Il s'agit d'agents de surface ou de compositions en comportant, tels que: savons, alkylsulfates, alkylsulfonates, condensats d'acides gras, alkylarylsulfonates, polyglycolesters et éthers.

de Waschmittel für die Textilindustrie
Produkt, das in der Textilindustrie zur Entfernung von Fett und Schmutzstoffen von Textilien während der Her-

stellung und Veredlung dient. Es ist dem jeweiligen Arbeitsgang angepasst und findet z.B. bei der Rohwollwäsche, bei der Garn- und Stückwäsche, beim Nachwaschen von Färbungen und Drucken Verwendung.
Anmerkung: Es handelt sich um grenzflächenaktive Stoffe oder Zubereitungen hieraus, die in der Regel als Seifen, Alkylsulfate, Alkylsulfonate, Fettsäurekondensationsprodukte, Alkylarylsulfonate, Polyglykolester und -äther vorliegen.

el agente detergente para la industria textil
Producto que sirve en la industria textil para eliminar las grasas y suciedad de los artículos textiles durante la fabricación y el acabado. Su composición y formulación debe estar adaptada a cada fase de trabajo, por ejemplo, al lavado de la lana bruta, de los hilados y de los tejidos en pieza, al lavado posterior de tintura y estampaciones, etc.
Observación: Se trata de agentes de superficie o de composiciones que los contienen, tales como jabones, alquilsulfatos, alquilsulfonatos, productos de condensación de ácidos grasos, alquilarilsulfonatos, ésteres y éteres de poliglicoles.

it —

ne wasmiddel voor de textielindustrie
Een produkt dat in de textielindustrie wordt gebruikt in de diverse produktiestadia om spin- of eindavivages of verontreinigingen te verwijderen. Zijn samenstelling komt overeen met de eisen voor de verschillende stappen van het proces, zoals het wassen van ruwe wol, garens en stukgoederen, voor het na-wassen van geverfde en bedrukte stoffen enz.

pl środek piorący dla przemysłu włókienniczego
Produkt stosowany w przemyśle włókienniczym do usuwania tłuszczu i brudu z tkanin podczas ich produkcji i wykańczania. Jego skład (receptura) dostosowany jest do różnych operacji technologicznych takich np. jak pranie surowej wełny, pranie przędzy lub sztuk tkanin, spieranie tkanin barwionych lub drukowanych.
Uwaga: Produktami takimi są związki powierzchniowo czynne takie jak mydła, alkilosiarczany, alkilosulfoniany, produkty kondensacji kwasów tłuszczowych, estry i etery poliglikoli oraz preparaty zawierające te związki.

185 detergent oil
A lubricating compound, generally based on mineral oils and surface active agents,

which facilitates the suspension or resuspension of solid particles emanating from the operation of an internal combustion engine.

fr huile détergente
Composition lubrifiante, généralement à base d'huiles minérales et d'agents de surface, facilitant la mise ou la remise en suspension des particules solides résultant du fonctionnement d'un moteur à combustion interne.

de Spülöl
Schmierstoffhaltige Mischung, im allgemeinen auf der Basis von Mineralöl und grenzflächenaktiven Substanzen, welche die Suspendierung und Wiedersuspendierung fester Teilchen, die beim Betrieb eines Motors auftreten, erleichtert.

el aceite detergente
Composición lubrificante, generalmente a base de aceites minerales y de agentes de superficie, que facilita la puesta en suspensión de las partículas sólidas producidas como resultado del funcionamiento de un motor de combustión.

it —

ne spoelolie
Een smeermiddel, in het algemeen gebaseerd op minerale olie en oppervlakaktieve stoffen, dat het in suspensie brengen van vaste deeltjes bevordert, die afkomstig zjin van een in werking zijnde verbrandingsmotor.

pl olej smarny z dodatkiem środków myjąco-dyspergujących
Produkt o działaniu smarnym, na ogół oparty na olejach mineralnych i środkach powierzchniowo czynnych, który ułatwia dyspergowanie i utrzymanie w zawiesinie stałych cząstek powstających w czasie pracy silników spalinowych.

186 detergent power
fr pouvoir détergent
de Waschkraft
el poder detergente, poder detersivo
it potere detergente
ne waskracht
pl zdolność piorąca

187 determination of detergency
fr mesure du pouvoir détergent
de Waschkraftbestimmung
el medida del poder detergente, determinación de la detergencia
it determinazione del potere detergente

ne bepaling van de waskracht
pl oznaczanie zdolności piorącej

188 determination of foaming power, determination of lathering power
fr mesure du pouvoir moussant, détermination du pouvoir moussant
de Schaumkraftmessung, Schaumbestimmung
el medida del poder espumante, determinación del poder espumante
it determinazione del potere schiumogeno
ne bepaling van het schuimvermogen
pl oznaczanie zdolności pienienia się

189 detersion, cleaning
The act of bringing the phenomenon of detergency into effect.
fr détersion
Action correspondant à la mise en œuvre du phénomène de détergence.
de Detersion, Reinigung
Ausführung der zum Ablauf des Waschvorgangs erforderlichen Massnahmen.
el detersión
Acción y efecto del proceso de la detergencia.
it detersione
Azione che fa sì che si esplichi un processo di detergenza.
ne wassen
Handeling die erop gericht is het verschijnsel waskracht een rol te doen spelen.
pl detersja, usuwanie zanieczyszczeń

190 dilatancy
Under isothermal and reversible conditions, increase without hysteresis of the apparent viscosity under a shearing load.
fr dilatance
Dans des conditions isothermes et réversibles, augmentation sans hystérésis de la viscosité apparente sous l'effet d'une contrainte de cisaillement croissante.
de Dilatanz
Unter isothermen und reversiblen Bedingungen, Vergrösserung (ohne Hysterese) der scheinbaren Viskosität unter dem Einfluss einer mechanischen Schubspannung.
el dilatancia
En condiciones isotermas y reversibles, aumento sin histéresis de la viscosidad aparente, por la acción de un esfuerzo de cizallamiento creciente.

it dilatanza
ne dilatantie
Onder isotherme en reversibele omstandigheden, zonder hysterese, verhoging van de schijnbare viscositeit onder de invloed van een mechanische afschuifspanning.
pl dylatancja
Podwyższenie pozornej lepkości (bez histerezy) pod wpływem działania mechanicznych sił ścinających w warunkach izotermicznych i odwracalnych.

191 direct dyestuff
fr colorant direct
de Direktfarbstoff
el colorante directo
it colorante diretto
ne direkte kleurstof
pl barwnik bezpośredni

192 discharging assistant†
Product which, when added to a printing paste of discharge, makes it po ssible to discharge colour satisfactorily in the case of a dye which is difficult to discharge.
Note: These products are mainly based on onium derivatives and ethoxylated amines.
fr adjuvant de rongeage†
Produit qui, ajouté à une pâte d'impression de rongeage, permet d'obtenir un enlevage satisfaisant dans le cas d'une teinture difficile à ronger.
Nota: Il s'agit essentiellement de produits à base de dérivés d'onium et d'amines éthoxylées.
de Ätzhilfsmittel
Produkt, das Druckpasten zugesetzt, bei schwer ätzbaren Färbungen einen ausreichenden Ätzeffekt ermöglicht.
Anmerkung: Es handelt sich im wesentlichen um Produkte auf der Basis von Oniumverbindungen und äthoxylierten Aminen.
el auxiliar de corrosión
Producto que añadido a una pasta de estampación, permite obtener un suficiente efecto de corrosión, en el caso de colorantes difíciles de corroer.
Observación: Se trata esencialmente de productos a base de compuestos onio y de aminas oxietiladas.
it —
ne beitsmiddel
Een produkt dat wordt aangebracht op de vezel om met een (beits-)kleurstof een complex (kleurstoflak) te vormen, dat beter aan de vezel gebonden is dan de kleurstof alleen.

pl wspomagacz wywabiania
Produkt dodawany do past drukarskich, umożliwiający uzyskanie dobrego wywabu przy trudno wywabiających się wybarwieniach.
Uwaga: Produkty tego typu oparte są głównie na związkach oniowych i oksyetylenowanych aminach.
† See appendix
Voir appendice

193 discolour, fade
fr décolorer
de verbleichen
el descolorar, desteñir
it decolorare, decolorire
ne verbleken
pl blaknąć, wypełzać, płowieć

194 dishwashing agent, dishwasher
fr détersif pour la vaisselle
de Tellerwaschmittel, Tellerspülmittel, Geschirrwaschmittel
el detergente para la vajilla
it detergente per stoviglie
ne vaatwasmiddel
pl środek do mycia naczyń

195 dishwashing effectiveness
fr action détergente sur les assiettes, pouvoir nettoyant sur la vaisselle
de Tellerwaschwirkung
el acción detergente sobre la vajilla, poder limpiador sobre la vajilla
it azione detergente sui piatti
ne waswerking op borden
pl skuteczność mycia naczyń

196 dishwashing trial, dishwashing test
fr essai de lavage des assiettes
de Tellerwaschversuch
el ensayo del lavado de vajilla
it prova di lavaggio dei piatti, prova di lavaggio delle stoviglie
ne bordenwasproef
pl test talerzowy, próba mycia naczyń

197 disinfectant
fr désinfectant
de Desinfektionsmittel
el desinfectante
it disinfettante
ne desinfectiemiddel
pl środek odkażający, środek dezynfekujący

198 dispenser
fr distributeur
de Spender

el distribuidor
it distributore
ne verdeler
pl dystrybutor

199 disperse phase
The discontinuous phase of a dispersion.
fr phase dispersée
Phase discontinue d'une dispersion.
de dispergierte Phase
Diskontinuierliche Phase einer Dispersion.
el fase dispersa
Fase discontinua de una dispersión.
it fase dispersa
Fase discontinua di una dispersione.
ne gedispergeerde fase
De diskontinue fase van een dispersie.
pl faza rozproszona, faza zdyspergowana
Nieciągła faza dyspersji.

200 dispersibility
fr dispersibilité
de Verteilbarkeit
el dispersibilidad
it dispersibilità
ne colloïdale verdeelbaarheid
pl podatność na dyspergowanie

201 dispersing agent
A substance capable of promoting the formation of a dispersion.
fr agent dispersant, dispersant
Produit apte à promouvoir la formation d'une dispersion.
de Dispergiermittel
Produkt, das in der Lage ist, die Bildung einer Dispersion zu fördern.
el agente dispersante, dispersante
Producto capaz de promover la formación de una dispersión.
it disperdente
Sostanza atta a promuovere la formazione di una dispersione.
ne dispergeermiddel
Een stof, die het vormen van een dispersie bevordert.
pl środek dyspergujący, dyspergator
Substancja zdolna do tworzenia dyspersji.

202 dispersing power
The extent of the ability of a product to bring about the formation of a dispersion.
fr pouvoir dispersant
Degré d'aptitude d'un produit à provoquer une dispersion.

de Dispergiervermögen
Grad der Fähigkeit eines Produktes, die Bildung einer Dispersion zu begünstigen.
el poder dispersante
Capacidad de una substancia para promover la formación de una dispersión.
it potere disperdente
Grado dell'attitudine di certe sostanze a formare una dispersione.
ne dispergerend vermogen
Het vermogen van een produkt een dispersie tot stand te brengen.
pl zdolność dyspergowania
Zdolność danego produktu do wytwarzania dyspersji.

203 dispersion
A system consisting of several phases of which one is continuous and at least one other is finely dispersed.
fr dispersion
Système de plusieurs phases dont l'une est continue et dont une autre au moins est finement répartie.
de Dispersion
System mehrerer Phasen, von denen eine kontinuierlich und mindestens eine andere fein verteilt ist.
el dispersión
Sistema de varias fases, de las cuales una es continua y por lo menos otra está finalmente repartida.
it dispersione
Sistema polifasico comprendente una fase continua e almeno una fase finemente suddivisa.
ne dispersie
Een systeem bestaande uit verschillende fasen, waarvan er één kontinu en ten minste één andere fijn gedispergeerd is.
pl dyspersja
Układ składający się z szeregu faz, z których jedna jest fazą ciągłą a co najmniej jedna z pozostałych jest fazą silnie rozproszoną.

204 dispersion medium
The continuous phase of a dispersion.
fr milieu de dispersion
Phase continue d'une dispersion.
de Dispersionsmedium
Kontinuierliche Phase einer Dispersion.
el medio de dispersión
Fase continua de una dispersión.

it mezzo di dispersione
    Fase continua di una dispersione.
ne dispersiemedium
    Kontinue fase van een dispersie.
pl faza rozpraszająca, ośrodek dyspergujący
    Faza ciągła dyspersji.

205 **disulphide bond**
fr liaison disulfure
de Disulfidbindung
el enlace disulfuro
it legame bisolfurico
ne disulfide-binding
pl wiązanie dwusiarczkowe

206 **disulphide bridge, cystine bridge**
fr pont disulfure, pont cystinique
de Disulfidbrücke, Cystinbrücke
el puente disulfuro, puente cisteínico
it ponte bisolfurico, ponte cistinico
ne disulfide-brug
pl mostek dwusiarczkowy, mostek
    cystynowy

207 **dithioglycolic acid**
fr acide dithioglycolique
de Dithioglykolsäure
el ácido ditioglicólico
it acido ditioglicolico
ne dithioglycolzuur
pl kwas dwutioglikolowy

208 **domestic hard soap**
fr savon de ménage, savon de Marseille
de Haushaltseife
el jabón casero, jabón de cocina, jabón
    corriente, jabón de Marsella
it sapone da bucato, sapone di Marsiglia
ne huishoudzeep
pl mydło gospodarcze

209 **domestic washing machine**
fr machine à laver ménagère, machine à
    laver domestique
de Haushaltwaschmaschine
el máquina de lavar doméstica, lavadora
    doméstica
it macchina lavatrice per la casa
ne huishoudwasmachine
pl pralka domowa

210 **dry shampoo**
fr shampooing sec
de Trockenshampoo
el champu seco
it shampoo secco
ne droge shampoo
pl szampon suchy

211 **drying**
fr séchage
de Trocknung
el secado
it essiccamento
ne drogen
pl suszenie

212 **dye-fixing agent**
    Product intended to improve the fast-
    ness of dyes from certain points of view.
    To increase the rubbing fastness, use
    is made of detergents for the textile
    industry which eliminate the loose dye.
    To increase the wet fastness, use is
    made of products which, together with
    the dye, form stable compounds which
    do not dissolve easily.
    Note: In the latter case, these products include
    cationic substances such as: amines and amine
    derivatives, for example, onium salts and oxal-
    kylated products.
fr agent pour le traitement subséquent des
    teintures
    Produit destiné à améliorer certaines
    solidités des teintures. Pour augmenter
    la solidité au frottement on utilise des
    agents détergents pour l'industrie textile
    éliminant le colorant non fixé. Pour aug-
    menter la solidité au mouillé on utilise
    des produits qui forment avec le colorant
    des composés stables difficilement so-
    lubles.
    Nota: Dans ce dernier cas, parmi ces produits
    on peut citer des substances cationiques telles
    que: amines et dérivés d'amines, par exemple,
    des sels d'onium et des produits oxyalkylés.
de Nachbehandlungsmittel für Färbungen
    Produkt, das dazu bestimmt ist, ge-
    wisse Echtheiten von Färbungen zu
    verbessern. Zur Verbesserung der Reib-
    echtheit arbeitet man mit waschaktiven
    Stoffen, die den nicht fixierten Farbstoff
    beseitigen. Zur Erhöhung der Nassecht-
    heit können Produkte dienen, die schwer
    lösliche, stabile Verbindungen mit den
    Farbstoffen bilden.
    Anmerkung: Im letztgenannten Falle handelt
    es sich bei diesen Produkten u.a. um kationaktive
    Stoffe, wie Amine und Aminderivate z.B. Onium-
    salze und Oxalkylierungsprodukte.
el agente para el tratamiento posterior de
    tinturas
    Producto destinado a mejorar algunas
    solideces de las tinturas. Para aumentar
    la solidez al frote, se utilizan agentes
    detergentes, que eliminan el colorante
    no fijado. Para aumentar la solidez

al mojado, se utilizan productos que
forman con el colorante compuestos
estables, difícilmente solubles.

Observación: Entre estos últimos productos
pueden citarse substancias catiónicas, como aminas y derivados de aminas, sales de compuestos
onio y aminas oxietiladas.

it —

ne verffixeermiddel
Een produkt dat de echtheid van een
verving in verschillende opzichten moet
verbeteren. Meestal gaat het hier om de
zgn. natechtheden.

pl środek pomocniczy do utrwalania wybarwień
Produkt, którego zadaniem jest poprawa
trwałości wybarwień. Dla poprawy od-
porności na tarcie wykorzystuje się
środki piorące, które usuwają nie
utrwalony barwnik. Dla poprawy od-
porności wybarwień na mokro służyć
mogą produkty tworzące z barwnikami
trudno rozpuszczalne, trwałe związki.

Uwaga: W tym ostatnim przypadku omawiane
produkty zawierają związki kationowe takie jak
aminy i pochodne amin, np. sole oniowe oraz
produkty oksyalkilenowane.

213 dyeing bath
fr bain de teinture
de Färbeflotte
el baño para teñir
it bagno di tintura
ne verfbad
pl kąpiel barwiąca, kąpiel farbiarska

214 dyeing procedure
fr procédé de teinture
de Färbeverfahren
el procedimiento de teñir
it processo di tintura
ne werkwijze van het verven
pl metoda barwienia, metoda farbowania,
proces barwienia

215 dyestuff
fr matière colorante
de Farbstoff
el colorante
it colorante
ne kleurstof
pl barwnik

E

216 effectiveness of the optical bleach
fr effet d'azurage optique
de optischer Aufhell-Effekt
el efectividad del blanqueo óptico
it effetto di sbiancaggio ottico
ne effekt van het optisch bleken
pl efekt rozjaśniania optycznego

217 elasticity test
fr essai d'extensibilité, épreuve d'extensibilité
de Dehnbarkeitsprobe
el prueba de la elasticidad, prueba de la extensibilidad
it saggio di elasticità
ne rekproef
pl próba na wydłużenie, badanie elastyczności, badanie rozciągliwości

218 electrical double layer
fr couche double électrique
de elektrische Doppelschicht
el doble capa eléctrica
it doppio strato elettrico
ne electrische dubbellaag
pl podwójna warstwa elektryczna, podwójna warstwa jonowa

219 emulsifiable liquid
A liquid suitable for constituting the disperse phase of an emulsion.
fr liquide émulsionnable
Liquide possédant une aptitude à constituer la phase discontinue d'une émulsion.
de emulgierbare Flüssigkeit
Flüssigkeit, die in der Lage ist, die diskontinuierliche Phase einer Emulsion zu bilden.
el líquido emulsionable
Líquido que posee la facultad de poder constituir la fase discontinua de una emulsión.
it liquido emulsionabile
Liquido atto a costituire la fase discontinua di una emulsione.
ne emulgeerbare vloeistof
Een vloeistof geschikt voor het vormen van de gedispergeerde fase van een emulsie.
pl ciecz emulgująca się
Ciecz, która może tworzyć fazę zdyspergowaną emulsji.

220 emulsification
The action causing the formation of an emulsion.
fr emulsification
Action entraînant la formation d'une émulsion.
de Emulgierung
Vorgang, der zu Bildung einer Emulsion führt.
el emulsionamiento
Acción que conduce a la formación de una emulsión.
it emulsificazione
Operazione che porta alla formazione di una emulsione.
ne emulgeren
De werking die de vorming van een emulsie veroorzaakt.
pl emulgowanie
Działanie powodujące powstawanie emulsji.

221 emulsifying agent, emulsifier†
A substance which permits or facilitates the formation of an emulsion.
fr agent émulsionnant, émulsifiant
Produit qui permet ou facilite la formation d'une émulsion.
de Emulgator
Produkt, das die Bildung einer Emulsion ermöglicht oder erleichtert.
el producto emulsionante, emulsionante
Producto que permite o facilita la formación de una emulsión.
it emulsionante
Sostanza che provoca o facilità la formazione di una emulsione.
ne emulgeermiddel
Een stof die het vormen van een emulsie mogelijk maakt of bevordert.
pl emulgator
Substancja umożliwiająca lub ułatwiająca powstawanie emulsji.
† See appendix

222 emulsifying agent for the textile industry
Product which makes possible or facilitates the formation of an emulsion. In the textile industry, it is generally used in the preparation of batching, brightening and/or preparing agents, winding oils, etc., in order to obtain a special effect.
Note: These products are surface active agents

or preparations comprising them, such as: soaps, alkylsulphates, alkylsulphonates, fatty acid condensates, alkylarylsulphonates, polyglycol esters and ethers, esters of fatty acids and polyhydroxyl compounds.

fr agent émulsionnant pour l'industrie textile

Produit qui permet ou facilite la formation d'une émulsion. Dans l'industrie textile, il est utilisé généralement dans la mise en œuvre des agents d'ensimage, d'avivage et/ou de préparation, des huiles de bobinage, etc., afin d'obtenir un effet spécial.

Nota: Il s'agit d'agents de surface ou de préparations en comportant, tels que: savons, alkylsulfates, alkylsulfonates, condensats d'acides gras, alkylarylsulfonates, esters et éthers de polyglycols, esters d'acides gras et de composés polyhydroxylés.

de Emulgiermittel für die Textilindustrie

Produkt, das die Bildung einer Emulsion ermöglicht oder erleichtert. In der Textilindustrie wird es in der Regel bei der Verarbeitung von Schmälzmitteln, Avivage- und/oder Präparationsmitteln, Spulölen usw. angewendet, um einen speziellen Effekt zu erzielen.

Anmerkung: Es handelt sich um grenzflächenaktive Stoffe oder Zubereitungen aus diesen, wie Seifen, Alkylsulfate, Alkylsulfonate, Fettsäurekondensationsprodukte, Alkylarylsulfonate, Polyglykolester und -äther, sowie Fettsäureester von Polyhydroxyverbindungen.

el agente emulsionante para la industria textil

Producto que permite o facilita la formación de una emulsión. En la industria textil se utilizan generalmente junto con agentes de ensimaje, de avivado y/o de preparación, con aceites de bobinado, etc., a fin de obtener un efecto especial.

Observación: Se trata de agentes de superficie o de preparaciones que los contienen, tales como jabones, alquilsulfatos, alquilsulfonatos, productos de condensación de ácidos grasos, alquilarilsulfonatos, ésteres y éteres de poliglicoles, así como ésteres de ácidos grasos y de compuestos polihidroxilados.

it —

ne emulgator voor de textielindustrie

Produkt dat de emulsievorming mogelijk of gemakkelijker maakt. Het wordt in de textielindustrie gewoonlijk gebruikt bij de bereiding van olietoevoegingsmiddelen (batching agents), optische bleekmiddelen en/of preparing agents, spoeloliën enz. voor het verkrijgen van een speciaal effect.

Opmerking: Deze produkten zijn oppervlakaktieve stoffen of preparaten die deze bevatten zoals zepen, alkylsulfaten, alkylsulfonaten, vetzuurcondensaten, alkylarylsulfonaten, polyglycolesters en -ethers, esters van vetzuren en polyalcoholen.

pl emulgator dla przemysłu włókienniczego

Produkt umożliwiający lub ułatwiający powstanie emulsji. W przemyśle włókienniczym stosowany jest on głównie dla uzyskania określonych efektów przy przygotowywaniu natłustek, środków ożywiających i/lub preparujących, olejów przewijarkowych (przędzalniczych) itp.

Uwaga: Produktami takimi są związki powierzchniowo czynne takie jak mydła, alkilosiarczany, alkilosulfoniany, produkty kondensacji kwasów tłuszczowych, alkiloarylosulfoniany, estry i etery poliglikoli, estry kwasów tłuszczowych, związki wielowodorotlenowe, oraz preparaty zawierające te związki.

223 emulsifying liquid

A liquid suitable for constituting the continuous phase of an emulsion.

fr liquide émulsionnant

Liquide possédant une aptitude à constituer la phase continue d'une émulsion.

de emulgierende Flüssigkeit

Flüssigkeit, die in der Lage ist, die kontinuierliche Phase einer Emulsion zu bilden.

el líquido emulsionante

Líquido que posee la facultad de poder constituir la fase continua de una emulsión.

it liquido emulsionante

Liquido atto a costituire la fase continua di una emulsione.

ne emulgerende vloeistof

Een vloeistof geschikt voor het vormen van de kontinue fase van een emulsie.

pl ciecz emulgująca

Ciecz, która może tworzyć fazę ciągłą emulsji.

224 emulsifying power

The ability of a substance to facilitate the formation of an emulsion.

fr pouvoir émulsionnant, pouvoir émulsifiant

Degré d'aptitude d'un produit à faciliter la formation d'une émulsion.

de Emulgiervermögen

Grad der Fähigkeit eines Produktes, die Bildung einer Emulsion zu erleichtern.

el poder emulsionante

Capacidad de un producto para facilitar la formación de una emulsión.

it potere emulsionante
Grado dell'attitudine di un prodotto a promuovere la formazione di una emulsione.

ne emulgerend vermogen
Het vermogen van een stof de vorming van een emulsie te bevorderen.

pl zdolność emulgowania
Zdolność danego produktu do ułatwiania powstawania emulsji.

225 emulsion
A heterogeneous system made by dispersing small globules of one liquid in another liquid which forms a continuous phase.

fr émulsion
Système hétérogène constitué par la dispersion de fins globules d'un liquide dans un autre liquide formant une phase continue.

de Emulsion
Heterogenes System, bestehend aus einer Dispersion feiner Tröpfchen einer Flüssigkeit in einer anderen, die kontinuierliche Phase bildenden Flüssigkeit.

el emulsión
Sistema heterogéneo constituido por la dispersión de pequeños glóbulos de un líquido en otro líquido que forma una fase continua.

it emulsione
Sistema eterogeneo prodotto per dispersione di goccioline di un liquido in un altro liquido costituente la fase continua.

ne emulsie
Een heterogeen systeem bereid door het verdelen van kleine bolletjes van een vloeistof in een andere vloeistof, die een kontinue fase vormt.

pl emulsja
Układ niejednorodny utworzony przez zdyspergowanie małych cząstek jednej cieczy w drugiej cieczy tworzącej fazę ciągłą.

226 emulsion stability
The ability of an emulsion to persist.

fr stabilité d'émulsion
Degré d'aptitude d'une émulsion à la persistance.

de Emulsionsstabilität
Grad der Fähigkeit einer Emulsion, beständig zu bleiben.

el estabilidad de emulsión
Aptitud de una emulsión para la persistencia.

it stabilità di una emulsione
Attitudine di una emulsione alla persistenza.

ne emulsiestabiliteit
Het vermogen van een emulsie zich te handhaven.

pl trwałość emulsji
Zdolność emulsji do nieulegania zmianom.

227 enamel
fr vernis
de Glasur
el esmalte
it vernice
ne vernis
pl emalia, szkliwo

228 enamel decalcification
fr décalcification de l'émail
de Entkalkung des Zahnschmelzes
el descalcificación del esmalte
it decalcificazione dello smalto
ne ontkalking van het tandemail
pl odwapnienie szkliwa, odwapnienie emalii

229 enamel remover
fr dissolvant (à vernis)
de Glasurentferner
el disolvente del barniz
it solvente dello smalto
ne vernisverwijderer
pl zmywacz do paznokci

230 endophily
Constitutional property which denotes the tendency of the whole or a part of a molecule to penetrate into or remain in a phase. It is characterised, with regard to the functional groups of the molecule, by the fact that such a group gives rise to a difference by diminution, of the variation of the chemical potential, when the molecules of the product pass from a gaseous ideal state to the phase under consideration.
Note: The value of the diminution of the variation in chemical potential, which results from the introduction of the functional group, is a function of concentration and temperature. Such a group can, depending on these variables, have either an endophilic or an exophilic character.

fr endophilie
Propriété constitutionnelle qui corres-

pond à la tendance de tout ou partie d'une molécule, à pénétrer ou à rester dans une phase. Elle se caractérise, par rapport aux groupements fonctionnels de la molécule, par le fait qu'un tel groupement contribue à une différence de variation en diminution, du potentiel chimique lorsque les molécules du produit passent de l'état gazeux idéal en la phase considérée.

Nota: La valeur de la diminution de variation du potentiel chimique, qui résulte de l'introduction du groupement fonctionnel, dépend de la concentration et de la température. Un tel groupement peut donc, selon ces variables, se présenter comme ayant un caractère endophile ou exophile.

de Endophilie

Konstitutionelle Eigenschaft eines Moleküls oder einer molekularen Gruppe, die der Tendenz entspricht, in eine Phase einzudringen oder darin zu bleiben. Sie ist bei den molekularen Gruppen charakterisiert durch eine Abnahme der Veränderung des chemischen Potentials beim eventuellen Übergang der Substanz, die aus den betreffenden Molekülen besteht, vom Zustand eines verdünnten Gases in die betrachtete Phase.

Anmerkung: Der Wert der Abnahme der Veränderung des chemischen Potentials beim Eindringen der molekularen Gruppe, ist eine Funktion der Konzentration und der Temperatur. Eine solche Gruppe kann daher, in Abhängigkeit von diesen Bedingungen, einen endophilen oder einen exophilen Charakter haben.

el endofilia

Propiedad constitucional de una substancia que corresponde a la tendencia de sus moléculas, o de alguno de sus grupos moleculares, a penetrar o a permanecer en una fase. Con respecto a los grupos funcionales de la molécula, se caracteriza por el hecho de que tales grupos contribuyen a una disminución de variación del potencial químico, cuando las moléculas de dicha substancia pasan desde el estado de gas diluido hasta la fase considerada.

Observación: El valor de la disminución de variación del potencial químico, debido a la introducción del grupo funcional depende de la concentración y de la temperatura. Según el valor de estas variables, ese grupo puede presentar un carácter endófilo o exófilo.

it endofilia

ne endofilie

Strukturele eigenschap die overeenkomt met de neiging van het gehele molecuul of een deel ervan, een fase

binnen te dringen of deze niet te verlaten. Met betrekking tot funktionele groepen van een molecuul wordt de eigenschap gekenmerkt door het feit dat een groep de neiging heeft, een verschil in variatie te veroorzaken door verlaging van de chemische potentiaal als gevolg van de mogelijke overgang van de door de moleculen van het produkt gevormde stof uit ideale gasvormige toestand naar de betreffende fase.

Opmerking: De mate van de verlaging van de chemische potentiaal als gevolg van de moleculaire groep is afhankelijk van de concentratie en de temperatuur. Een dergelijke groep kan, afhankelijk van deze variabelen, een endofiel of een exofiel karakter hebben.

pl endofilia

Właściwość strukturalna, określająca skłonność danej cząsteczki lub jej części do wnikania w głąb danej fazy lub do pozostawania wewnątrz niej. W odniesieniu do grup funkcyjnych cząsteczki charakteryzuje się tym, że grupa taka powoduje zmniejszenie zmian potencjału chemicznego przy przejściu cząsteczek substancji ze stanu gazu doskonałego do rozpatrywanej fazy.

Uwaga: Wartość zmniejszenia zmiany potencjału chemicznego, która wynika z wprowadzenia grupy funkcyjnej, jest funkcją stężenia i temperatury. Grupa taka, w zależności od tych zmiennych, może więc posiadać charakter endofilny lub egzofilny.

231 epidermis
fr épiderme
de Epidermis
el epidermis
it epidermide
ne opperhuid
pl naskórek

232 esterification
In the particular case of surface active agents, chemical reaction giving rise to an ester derived from an acid and an alcohol, enol or phenol with the elimination of water.

fr estérification
Dans le cas particulier des agents de surface, réaction chimique permettant d'obtenir un ester à partir d'un acide et d'un alcool, énol ou phénol, avec élimination d'eau.

de Veresterung
Im speziellen Fall der grenzflächenaktiven Körper: chemische Reaktion, bei der ein Ester gebildet wird aus einer

Säure und einem Alkohol, Enol oder
Phenol unter Abspaltung von Wasser.

el esterificación
En el caso particular de los agentes
de superficie: reacción química que
permite obtener un éster a partir de
un ácido y de un alcohol, enol o fenol,
con eliminación de agua.

it esterificazione
Nel caso particolare dei tensioattivi:
reazione chimica che porta alla formazione di un estere da acido e alcole, enolico
o fenolico, con eliminazione di acqua.

ne verestering
In het speciale geval van oppervlakaktieve stoffen is dit een chemische
reactie waarbij een ester ontstaat uit
een zuur en een alcohol, enol of fenol
onder afgifte van water.

pl estryfikacja
W przypadku związków powierzchniowo
czynnych reakcja chemiczna prowadząca
do utworzenia z kwasu oraz alkoholu,
enolu lub fenolu estrów z jednoczesnym
wydzieleniem wody.

233 ethoxylation
In the particular case of surface active
agents, chemical reaction leading to
the addition of one or more molecules
of ethylene oxide to a labile hydrogen
compound.

fr éthoxylation, éthoxylénation
Dans le cas particulier des agents de
surface, réaction chimique permettant
la fixation d'une ou plusieurs molécules
d'oxyde d'éthylène sur un composé
à hydrogène labile.

de Äthoxylierung
Im speziellen Fall der grenzflächenaktiven Körper: chemische Reaktion, bei
welcher ein oder mehrere Mol Äthy-
lenoxyd an eine Verbindung mit aktivem Wasserstoff angelagert werden.

el etoxilación, etoxilenación
En el caso particular de los agentes
de superficie: reacción química que
permite la fijación de una o de varias
moléculas de óxido de etileno, en un
compuesto con hidrógeno lábil.

it etossilazione
Nel caso particolare dei tensioattivi:
reazione chimica di addizione di una
o più molecole di ossido d'etilene su
un composto a idrogeno mobile.

ne ethoxylering
In het bijzondere geval van oppervlakaktieve stoffen is dit de chemische
reactie waarbij een of meer moleculen
ethyleenoxide geaddeerd worden aan een
labiele waterstofverbinding.

pl oksyetylenowanie
W przypadku związków powierzchniowo
czynnych reakcja chemiczna, w której
zachodzi przyłączenie jednej lub więcej
cząsteczek tlenku etylenu do związku
posiadającego ruchliwy wodór.

234 eutrophication
Proliferation and excessive growth of
algae and other aquatic plants, which
results from the very complex interaction of a number of factors.

fr eutrophisation
Prolifération et croissance excessive
d'algues et autres plantes aquatiques,
qui résultent de l'interaction très complexe de multiples facteurs.

de Eutrophierung
Ausbreitung und übermässiges Wachs-
tum von Algen und anderen Wasser-
pflanzen, welche durch die sehr komplexe
Wechselwirkung verschiedener Faktoren hervorgerufen werden.

el eutroficación
Proliferación y crecimiento excesivo
de algas y de o'ras plantas acuáticas,
que resulta de la interacción muy
compleja de múltiples factores.

it eutrofizzazione
Proliferazione e crescita eccessiva di
alghe e altre piante acquatiche dovuto
a interazione molto complessa di diversi fattori.

ne eutroficatie
Uitbreiding en overmatige groei van
algen en andere waterplanten, als gevolg van een zeer komplexe wissel-
werking van verschillende faktoren.

pl eutrofizacja
Rozmnażanie się i nadmierny wzrost
alg i innych roślin wodnych będący
wynikiem bardzo złożonego oddziaływania szeregu czynników.

235 evaluation of foaming properties
fr appréciation de la mousse
de Schaumbeurteilung
el determinación de la espuma
it valutazione della schiuma
ne beoordeling van het schuim

pl   ocena piany, ocena zdolności pienienia się

236 evenness
fr   uniformité
de   Egalisierwirkung
el   uniformidad
it   uniformità
ne   egaliserende werking
pl   wyrównanie, egalizacja

237 exophily
Constitutional property which denotes the tendency of the whole or a part of a molecule to pass out of or not to penetrate into a phase. It is characterised, with regard to the functional groups of the molecule, by the fact that such a group gives rise to a difference by increase of the variation of the chemical potential, when the molecules of the product pass from a gaseous ideal state to the phase under consideration.
Note: The value of the increase of the variation in chemical potential, which results from the introduction of the functional group, is a function of concentration and temperature. Such a group can, depending on these variables, have either an endophilic or an exophilic character.

fr   exophilie
Propriété constitutionnelle, qui correspond à la tendance de tout ou partie d'une molécule, à ne pas pénétrer dans une phase, ou à en sortir. Elle se caractérise, par rapport aux groupements fonctionnels de la molécule, par le fait qu'un tel groupement contribue à une différence de variation en augmentation du potentiel chimique lorsque les molécules du produit passent de l'état gazeux idéal en la phase considérée.
Nota: La valeur de l'augmentation de variation du potentiel chimique, qui résulte de l'introduction du groupement fonctionnel, dépend de la concentration et de la température. Un tel groupement peut donc, selon ces variables, se présenter comme ayant un caractère endophile ou exophile.

de   Exophilie
Konstitutionelle Eigenschaft eines Moleküls oder einer molekularen Gruppe, die der Tendenz entspricht, nicht in eine Phase einzudringen oder diese zu verlassen. Sie ist bei den molekularen Gruppen charakterisiert durch eine Zunahme der Veränderung des chemischen Potentials beim eventuellen Übergang der Substanz, die aus den betreffenden Molekülen besteht, aus dem verdünnten Gaszustand in die betrachtete Phase.
Anmerkung: Der Wert der Zunahme der Veränderung des chemischen Potentials beim Eindringen der molekularen Gruppe, ist eine Funktion der Konzentration und der Temperatur. Eine solche Gruppe kann daher, in Abhängigkeit von diesen Bedingungen, einen endophilen oder einen exophilen Charakter haben.

el   exofilia
Propiedad constitucional de la substancia que corresponde a la tendencia de sus moléculas o de alguno de sus grupos moleculares, a no penetrar en una fase o a salir de ella. Con respecto a los grupos funcionales de la molécula, se caracteriza por el hecho de que tales grupos contribuyen a un aumento de variación del potencial químico, cuando las moléculas de dicha substancia pasan desde el estado de gas diluido hasta la fase considerada.
Observación: El valor del aumento de la variación del potencial químico debido a la introducción del grupo funcional depende de la concentración y de la temperatura. Según el valor de estas variables, ese grupo puede presentar carácter endófilo o exófilo.

it   esofilia
ne   exofilie
Structurele eigenschap die overeenkomt met de neiging van het gehele molecuul of een deel ervan, een fase niet binnen te dringen of te verlaten. Met betrekking tot funktionele groepen van een molecuul wordt deze eigenschap gekenmerkt door het feit dat een groep de neiging heeft, een verschil in variatie te veroorzaken door verhoging van de chemische potentiaal als gevolg van de mogelijke overgang van de door de moleculen van het produkt gevormde stof, uit de ideale gasvormige toestand naar de betreffende fase.
Opmerking: De mate van de verhoging van de chemische potentiaal als gevolg van de invoering van de moleculaire groep is afhankelijk van de concentratie en de temperatuur. Een dergelijke groep kan, afhankelijk van deze variabelen, een endofiel of een exofiel karakter hebben.

pl   egzofilia
Właściwość strukturalna, określająca skłonność danej cząsteczki lub jej części do opuszczania danej fazy lub brak skłonności do wnikania w jej głąb. W odniesieniu do grup funkcyjnych cząsteczki charakteryzuje się tym, że grupa taka powoduje wzrost zmian potencjału chemicznego przy przejściu

cząsteczek substancji ze stanu gazu doskonałego do rozpatrywanej fazy.

Uwaga: Wartość przyrostu zmiany potencjału chemicznego, która wynika z wprowadzenia grupy funkcyjnej, jest funkcją stężenia i temperatury. Grupa taka, w zależności od tych zmiennych, może więc posiadać charakter endofilny lub egzofilny.

238 expectorant
fr expectorant
de schleimlösendes Mittel
el expectorante
it espettorante
ne slijmverwijderend middel
pl środek wykrztuśny

239 eye cream
fr crème pour les yeux
de Lidcreme
el crema para los ojos

it crema per gli occhi
ne ogencrème
pl krem do powiek

240 eye lotion
fr lotion pour les yeux
de Augenwasser
el loción para los ojos
it lozione per gli occhi
ne oogwater
pl płyn do oczu, krople do oczu

241 eye shadow
fr ombre pour les yeux
de Lidschatten
el sombra para los ojos
it ombretto
ne oogschaduw
pl kredka do powiek, cień do powiek

F

242 **face mask, facial mask**
fr masque pour le visage
de Gesichtspackung
el máscara facial
it maschera per il viso
ne gezichtsmasker
pl maseczka kosmetyczna,
maseczka do twarzy

243 **face powder**
fr poudre de beauté, poudre
de Gesichtspuder
el polvos faciales
it cipria
ne gezichtspoeder
pl puder do twarzy

244 **facial lotion**
fr lotion faciale, lotion pour le visage
de Gesichtswasser, flüssige Gesichtscreme
el loción facial
it lozione per il viso
ne gezichtswater
pl płyn do twarzy

245 **fashionable shade**
fr couleur à la mode, teinte à la mode
de modische Nuance
el color de moda, tono a la moda
it tinta di moda
ne modetint
pl modny odcień

246 **fast to wash**
fr solide au lavage
de waschecht
el resistente al lavado
it solido al lavaggio
ne wasbestendig
pl odporny na pranie

247 **fastness to light**
fr solidité à la lumière
de Lichtechtheit
el resistencia a la luz
it solidità alla luce
ne lichtbestendigheid
pl trwałość na światło, odporność na
światło

248 **feel**
fr toucher
de Griff
el tacto
it tatto

ne greep
pl chwyt, dotyk

249 **felting**
fr feutrage
de Filzen, Verfilzung, Filzbildung
el enfieltrado
it feltraggio
ne vervilten
pl spilśnianie, filcowanie

250 **fibre**
fr fibre
de Faser
el fibra
it fibra
ne vezel
pl włókno

251 **fibre humectant**
Product intended to control the desired
humidity of the yarns, to maintain it
throughout the subsequent textile operations, and possibly to increase the
strength of the yarns and their form
stability.
Note: These products are generally solutions of
wetting agents with hygroscopic agents and/or
preserving agents added.

fr agent pour l'humidification des filés
Produit destiné à régulariser l'humidité
désirée des filés, à la maintenir au
cours des opérations textiles ultérieures
et éventuellement à augmenter la résistance des filés et leur stabilité de
forme.
Nota: Il s'agit en général de solutions d'agents
mouillants additionnées d'agents hygroscopiques
et/ou d'agents de conservation.

de Garnbefeuchtungsmittel
Produkt, das dazu bestimmt ist, den
gewünschten Feuchtigkeitsgehalt in Gar-
nen einzustellen und im Verlaufe der
weiteren textilen Arbeitsgänge aufrecht
zu erhalten, sowie gegebenenfalls eine
Festigkeitszunahme oder Formstabilität
der Garne herbeizuführen.
Anmerkung: Es handelt sich im allgemeinen
um Netzmittellösungen denen hygroskopische
Mittel und/oder Konservierungsmittel zugesetzt
sind.

el agente de humidificación de los hilados
Agente destinado a regularizar el deseado grado de humedad de los hilados,
manteniéndolo constante en el curso

de las operaciones textiles posteriores, y en algunos casos, a aumentar la resistencia de las fibras.

Observación: Se trata en general, de disoluciones de agentes humectantes, adicionadas de agentes higroscópicos y/o de agentes de conservación.

it —

ne garenbevochtiger
Een produkt dat het vochtgehalte van het garen beïnvloedt.

pl środek do nawilżania przędzy
Produkt służący do doprowadzenia wilgotności przędzy do pożądanego poziomu i utrzymywania tej wilgotności podczas prowadzonych następnie operacji włókienniczych, jak również do ewentualnego podnoszenia wytrzymałości przędzy i trwałości jej kształtu.

Uwaga: Produkty te są zazwyczaj roztworami środków zwilżających z dodatkiem środków higroskopijnych i/lub środków konserwujących.

252 fibre incrustation
fr incrustation de la fibre
de Faserinkrustierung
el incrustación en la fibra
it incrostazione della fibra
ne incrustatie van de vezel
pl inkrustacja włókna

253 fibre protecting agent
Product used traditionally for preserving fibres, particularly animal fibres, during the operations of bleaching, dyeing and stripping.

Note: These products are based, for example, on degraded proteins, fatty acid and protein condensates, ammonium alkylsulphates and alkylsulphonates, salts of ligno-sulphonic acids.

fr agent de protection des fibres
Produit servant traditionnellement à ménager les fibres, en particulier animales, au cours des opérations de blanchiment, de teinture et de démontage.

Nota: Il s'agit de produits à base, par exemple, de protéines dégradées, de condensats d'acides gras et de protéines, d'alkylsulfates et d'alkylsulfonates d'ammonium, de sels d'acides lignosulfoniques.

de Faserschutzmittel
Produkt, das — historisch bedingt — zur Schonung insbesondere tierischer Fasern beim Bleichen, Färben und Abziehen eingesetzt wird.

Anmerkung: Es handelt sich um Produkte auf der Basis von z.B. Eiweissspaltprodukten, Fettsäure-Eiweisskondensationsprodukten, Ammoniumsalzen von Alkylsulfaten und Alkylsulfonaten und ligninsulfonsauren Salzen.

el agente protector de fibras
Producto que sirve para preservar las fibras, especialmente animales, en las operaciones de blanqueo, tintura y desmontado.

Observación: Se trata, por ejemplo, de productos a base de proteínas degradadas, de productos de condensación de ácidos grasos y proteínas, de sales amónicas de alquilsulfatos y de alquilsulfonatos y de sales de ácidos lignosulfónicos.

it —

ne vezelbeschermend middel
Een produkt dat gebruikt wordt voor het conserveren van vezels, vooral dierlijke, tijdens het bleek-, verf- en ontkleuringsproces.

Opmerking: Deze produkten zijn bijvoorbeeld gebaseerd op afgebroken eiwitten, vetzuur- en proteïnecondensaten, ammoniumalkylsulfaten en alkylsulfonaten, lignosulfonzure zouten.

pl środek ochraniający włókno
Produkt tradycyjnie stosowany do ochrony włókien, szczególnie pochodzenia zwierzęcego, podczas operacji bielenia, barwienia i odbarwiania.

Uwaga: Preparaty te otrzymywane są na bazie np. produktów degradacji białek, kondensatów białek z kwasami tłuszczowymi, alkilosiarczanów i alkilosulfonianów amonowych oraz soli kwasów ligninosulfonowych.

254 filler (for surface active agents)
Organic or inorganic product, usually inert, employed to produce the desired type of presentation and/or concentration. Examples: sodium sulphate, water, alcohol, etc.

fr charge (pour agents de surface)
Produit minéral ou organique, généralement inactif, servant à la mise au type de présentation et/ou de concentration d'un détergent. Exemples: sulfate de sodium, eau, alcool, etc.

de Füllstoff (für Tenside)
Anorganisches oder organisches, im allgemeinen inaktives Produkt, welches dazu dient, die Verkaufsform und/oder die Konzentration eines Waschmittels einzustellen. Beispiele: Natriumsulfat, Wasser, Alkohol, etc.

el carga (para agentes de superficie)
Producto mineral u orgánico, generalmente inactivo, que sirve para dar a un detergente la forma de presentación y/o la concentración deseadas. Ejemplos: sulfato sódico, agua, alcohol, etc.

it carica
Prodotto inorganico o organico, generalmente inerte, usato per ottenere il tipo desiderato di presentazione e/o

di concentrazione di un detergente. Esempi: solfato di sodio, acqua, alcole ecc.

ne vulstof (voor oppervlakaktieve stoffen)
Een organisch of anorganisch, meestal inert produkt dat gebruikt wordt voor het verkrijgen van het gewenste uiterlijk en/of concentratie. Voorbeelden: natriumsulfaat, water, alcohol.

pl wypełniacz, środek wypełniający (dla związków powierzchniowo czynnych)
Produkt organiczny lub nieorganiczny, na ogół nieczynny, dodawany dla uzyskania określonego wyglądu i/lub określonego stężenia. Przykłady: siarczan sodowy, woda, alkohol itd.

255 film
A thin layer of matter, homogeneous or not.

fr feuil
Mince épaisseur de matière, homogène ou non.

de Film, Schicht, Folie
Materie, welche in einer Dimension eine sehr geringe Ausdehnung hat. Sie kann homogen oder inhomogen sein.

el película, film
Materia en capa de muy fino espesor. Puede ser homogénea o heterogénea.

it pellicola
Strato sottile di materia omogenea o non omogenea.

ne film
Een dunne homogene of inhomogene laag van een stof.

pl warstewka, błona
Cienka, jednorodna lub niejednorodna warstewka substancji.

256 fine laundering
fr lavage délicat, blanchisserie fine, lessivage fin
de Feinwäsche
el lavado de prendas finas, lavado de prendas delicadas
it lavaggio di indumenti delicati, biancheria fine
ne fijne was
pl pranie delikatne

257 fingernail elongator
fr faux ongles
de Fingernagelverlängerer
el uñas postizas
it false unghie
ne nagelverlenger
pl sztuczne paznokcie

258 finished soap, soap on nigre
The phase of soap in the pan, anisotropic and translucent, smectic in structure, in equilibrium with the nigre. Finished soap constitutes the final pan product which rises on the nigre and which contains normally and fairly constantly 62–65% of total fatty acids, together with very small quantities of sodium hydroxide, chloride and glycerine.

fr savon lisse, savon liquidé, savon levé sur gras
Phase du savon en chaudière, anisotrope et translucide, à structure smectique, en équilibre avec le gras. Le savon lisse constitue le produit final de chaudière, lorsqu'on lève sur gras, et contient normalement, et de façon assez constante, 62 à 65% d'acides gras totaux, ainsi que de faibles quantités d'hydroxyde, de chlorure de sodium et de glycérine.

de geschliffener Seifenkern, Seifenkern auf Leimniederschlag
Anisotrope, durchscheinende Seifenphase smektischer Struktur, die im Gleichgewicht mit dem Leimniederschlag steht. Sie enthält normalerweise und fast konstant 62–65% Fettsäuren, sowie geringe Mengen freier Alkalien, Kochsalz und Glycerin.

el jabón liquidado, jabón liso
Fase anisótropa y translúcida del jabón, de estructura esméctica, con 62 a 65% de ácidos grasos, en equilibrio con los bajos. El jabón liquidado contiene pequeñas cantidades de álcali y de cloruro sódico, así como de glicerina si se ha partido de grasas.
Observación: Estado esméctico—estado de las fases cristalino-líquidas en el cual las moléculas o partículas anisótropas están regularmente orientadas en dos direcciones. En la dirección restante, las moléculas se orientan al azar.

it sapone liquidato, sapone finito
Fase del sapone in caldaia, a carattere anisotropo e di aspetto translucido di struttura pastosa, in equilibrio con le collette. Il sapone liquidato constituisce il prodotto finale di caldaia, che si trova sopra le collette e che contiene normalmente un prodotto a 62–65% di acidi grassi totali, insieme a parti molto

piccole di sodio idrato, sodio cloruro
e glicerina.

ne afgemaakte zeep
Anisotrope, doorschijnende zeepfase met
smektische struktuur, die in evenwicht
met de lijmzeep staat. Zij bevat meestal en
vrij konstant 62–65% vetzuren alsmede
kleine hoeveelheden natriumhydroxide,
chloride en glycerine.

pl wysół płynny, rdzeń
Anizotropowa, półprzezroczysta faza
mydła o smektycznej budowie, pozosta-
jąca w równowadze z klejem mydlanym.
Wysół płynny stanowi końcowy produkt warzenia mydła, znajdujący się na
kleju mydlanym, zawierający zazwyczaj
prawie stałą (62 do 65%) ilość kwasów
tłuszczowych oraz niewielkie ilości wol-
nych alkaliów, chlorków i gliceryny.

259 finishing
A final pan operation to bring the mass
to one of the following two or three-phase
equilibrium states, by boiling and addition of water and/or electrolytes:
— curd soap–lye (soap on lye)
— finished   soap–nigre–lye (invariant
system)
— finished soap–nigre (soap on nigre)
at the same time, imparting to the mass
a suitable viscosity for a good separation
of the phases. Finishing leads to the
production by decantation or centrifuging, of the phases curd soap or finished soap, of constant composition,
virtually free from impurities.

fr liquidation
Opération finale en chaudière, consistant à amener la masse par ébullition
et addition d'eau et/ou d'électrolytes,
à un des états d'équilibre à deux ou
trois phases suivants:
— savon grainé–lessive inférieure (savon levé sur lessive)
— savon lisse–gras–lessive inférieure
(système invariant)
— savon lisse–gras (savon levé sur gras)
tout en donnant à la masse une viscosité
convenable pour une bonne séparation
des phases. La liquidation permet
d'obtenir par décantation ou par centrifugation les phases savon grainé ou
savon lisse, à composition presque constante, avec élimination très poussée
des impuretés.

de Ausschleifen
Endbehandlung im Siedekessel, die darin
besteht, durch Zusatz von Wasser und
Elektrolyt, den Seifenkern in einen der
folgenden zwei- oder dreiphasigen Gleich-
gewichtszustände zu bringen:
— geronnener Kern–Unterlauge (Seifenkern auf Unterlauge)
— geschliffener Kern–Leimniederschlag–
Unterlauge (invariantes System)
— geschliffener Kern–Leimniederschlag
(Seifenkern auf Leimniederschlag).
In diesen erhält die Masse die nötige
Viskosität für eine gute Phasentrennung.
Durch anschliessendes Absitzenlassen
oder Ausschleudern erhält man einen
Seifenkern von fast konstanter Zusammensetzung.

el liquidación
Tratamiento final del jabón graneado
con agua y electrólitos, para llegar a un
estado de equilibrio, generalmente de
tres fases: "jabón graneado–bajos–lejía
inferior". Este tratamiento da a la masa
la viscosidad conveniente para una
buena separación de fases. Por sedimentación o centrifugación se obtiene
el jabón liquidado, de composición casi
constante.

it liquidazione
L'operazione finale effettuata in caldaia
per portare la massa ad uno dei reguenti
stati di equilibrio, a due o tre fasi, a
mezzo bollitura e aggiunta di acqua e/o
elettroliti:
— sapone levato–lisciva alcalina (sapone su lisciva)
— sapone finito–collette–lisciva alcalina (sistema a tre fasi)
— sapone finito–collette (sapone su
collette).
Il tutto, impartendo alla massa una
viscosità idonea, per una buona separazione delle fasi. La liquidazione porta
alla produzione per mezzo di decantazione o centrifugazione, delle fasi di
sapone levato, o di sapone finito, di
composizione costante praticamente pri-
vo di impurezze.

ne afmaken, likwideren, liquideren, slijpen
Eindbehandeling in de kookpan om door
toevoeging van water en electrolyt de
kernzeep in een der volgende twee- of
drie-fasenevenwichten te brengen:

— kernzeep–onderloog (kernzeep, op onderloog)
— afgemaakte kernzeep–lijmzeep–onderloog (invariant systeem)
— afgemaakte kernzeep–lijmzeep (zeepkern op lijmzeep).

pl wykańczanie
Operacja końcowa w kotle warzelnym polegająca na przeprowadzeniu masy mydlanej w jeden z poniższych dwu- lub trójfazowych stanów równowagowych przez domydlenie oraz dodatek wody i/lub elektrolitu:
— wysół ścięty–ług spodni (mydło na ługu spodnim)
— wysół płynny–klej mydlany–ług spodni (układ trójfazowy)
— wysół płynny–klej mydlany (mydło na kleju mydlanym)
w wyniku czego masa mydlana uzyskuje lepkość odpowiednią dla dobrego rozdzielenia się faz. Po dekantacji lub odwirowaniu otrzymuje się fazy wysołu ściętego lub wysołu płynnego o stałym składzie, praktycznie wolne od zanieczyszczeń.

260 finishing assistant
Product added to finishing compounds to impart fluidity, body and/or stability and to alter the finishing in the desired way.
Note: These products include, among others, sulphated oils and greases, and products mentioned under the heading of preparing agent.

fr adjuvant d'apprêtage
Produit ajouté aux compositions d'apprêts pour leur donner plus de fluidité, plus de corps et/ou plus de stabilité, ainsi que pour modifier dans un sens désiré l'effet d'apprêtage.
Nota: Il s'agit, entre autres, d'huiles et de graisses sulfatées, et de produits cités sous agent de préparation.

de Appreturhilfsmittel
Produkt, das Appreturmitteln zugesetzt wird, um diese zügiger, geschmeidiger und/oder stabiler zu machen, sowie um den Appretureffekt in einer gewünschten Richtung zu modifizieren.
Anmerkung: Es handelt sich unter anderem um sulfierte Öle und Fette und die unter Präparationsmittel genannten Produkte.

el auxiliar para el apresto
Producto añadido a las composiciones de apresto para darles más fluidez, suavidad y/o estabilidad, así como para mo-dificar en el sentido deseado el efecto de apresto.
Observación: Se trata, entre otros, de aceites y grasas sulfatados y de los pròductos citados en agente de preparación.

it —

ne appreteerhulpmiddel
Produkt dat aan apprets wordt toegevoegd om een grotere fluïditeit, body en/of stabiliteit te verkrijgen en om de appretuur in de gewenste richting te leiden.
Opmerking: Deze produkten omvatten o.a. gesulfateerde oliën en vetten en de produkten genoemd onder avivage.

pl środek pomocniczy przy wykańczaniu
Produkt dodawany do środków wykańczalniczych dla nadania im odpowiedniej płynności, konsystencji i/lub trwałości oraz dla zmodyfikowania w pożądanym kierunku samego procesu wykańczania.
Uwaga: Produkty te obejmują między innymi siarczanowane oleje i tłuszcze oraz produkty objęte definicją środków do preparacji.

261 fixation, neutralizing
fr fixation
de Fixierung
el fijación
it fissativo
ne fixatie
pl utrwalanie

262 flavour
fr arôme, parfum
de Aroma
el aroma, perfume
it aroma, profumo
ne smaakstof
pl smak, zapach, aromat

263 flocculate
Matter which has undergone flocculation.
fr floculat (floc)
Matière ayant subi la floculation.
de Flockulat
Substanz, welche der Flockung unterlegen ist.
el flóculos
Materia que ha sufrido la floculación.
it floculato
ne vlokken
pl flokulat
Substancja, która uległa flokulacji.

264 flocculation
fr floculation

de Flockung
el floculación
En una disolución coloidal, fenómeno más o menos reversible de formación de agregados, que por reposo tienen tendencia a sedimentar.
it floculazione
ne vlokking
pl flokulacja, kłaczkowanie

265 fluoride toothpaste
fr pâte dentifrice au fluor
de fluorhaltige Zahnpaste
el pasta dentífrica al fluor
it pasta dentifricia al fluoro
ne fluoride-tandpasta
pl fluorowa pasta do zębów

266 fly
fr ébouriffement
de Fliegen des Haares
el desgreñar
it arruffamento (di capelli)
ne verward zijn van het haar
pl rozsypywanie się włosów, puszystość włosów

267 foam
A mass of gas cells separated by thin films of liquid and formed by the juxtaposition of bubbles, giving a gas dispersed in a liquid.
fr mousse
Ensemble de cellules gazeuses séparées par des lames minces de liquide, et formé par la juxtaposition de bulles que donne un gaz dispersé dans un liquide.
de Schaum
Gesamtheit der durch dünne Flüssigkeitslamellen getrennten Zellen, gebildet durch Zusammenlagerung von Blasen, die durch ein in einer Flüssigkeit dispergiertes Gas erzeugt sind.
el espuma
Conjunto de células gaseosas, separadas por láminas delgadas de líquido, formado por la superposición de burbujas que forma un gas al dispersarse en un líquido.
it schiuma
Aggregato di cellule gassose separate da pellicole sottili di liquido e formato dalla giustapposizione di bolle, che dà un gas disperso in un liquido.
ne schuim
Een groot aantal kleine gasbellen gescheiden door dunne vloeistoffilms, ge-

vormd door de samenvoeging van bellen, wat als resultaat de dispersie van gas in een vloeistof heeft.
pl piana
Masa komórek gazowych oddzielonych cienkimi błonkami cieczy, utworzona w wyniku skupienia się obok siebie pęcherzyków otrzymanych poprzez zdyspergowanie gazu w cieczy.

268 foam booster
A product which increases the foaming power.
fr renforçateur de mousse
Produit augmentant le pouvoir moussant.
de Schaumverstärker
Substanz, welche das Schaumvermögen einer Lösung erhöht.
el reforzador de espuma
Producto que aumenta el poder espumante.
it rafforzatore di schiuma
ne schuimverbeteraar
Een stof die het schuimvermogen van een oplossing verhoogt.
pl wspomagacz pienienia
Produkt poprawiający własności pianotwórcze.

269 foam drainage
The return to the liquid phase of the excess of liquid entrained by bubbles during foaming.
fr essorage de la mousse
Retour à la phase liquide de l'excès du liquide entraîné par les bulles lors du moussage.
de Schaumentwässerung
Rückfluss der während des Schäumens von den Gasblasen in den Schaum eingebrachten Flüssigkeit zur flüssigen Phase.
el escurrido de la espuma
Retorno a la fase líquida del exceso de líquido arrastrado por las burbujas, durante la espumación.
it drenaggio della schiuma
Riflusso dell'eccesso di liquido trascinato dalle bolle durante lo schiumeggiamento.
ne drainage van het schuim
Terugkeer in de vloeistoffase van de overtollige vloeistof, die tijdens de schuimvorming werd meegevoerd.
pl odwadnianie piany, odciekanie piany
Powrót do fazy ciekłej nadmiaru cieczy porwanej przez pęcherzyki podczas pienienia.

270 **foam film**
fr lame de mousse
de Schaumlamelle
el capa de espuma, película de espuma
it pellicola della schiuma, film della schiuma
ne schuimvliesje
pl błonka piany

271 **foam height**
fr hauteur de mousse
de Schaumhöhe
el altura de la espuma
it altezza della schiuma
ne schuimhoogte
pl wysokość piany

272 **foam improver, lather booster, suds booster (USA)**
fr exalteur de mousse, agent améliorant de mousse, renforçateur de mousse
de Schaumverbesserer
el agente de mejoración de espuma, agente que aumenta la espuma
it esaltatore di schiuma
ne schuimverbeteraar
pl środek polepszający pienistość

273 **foam meter**
fr moussemètre
de Schaumapparat
el espumómetro
it misuratore di schiuma
ne schuimapparaat
pl przyrząd do pomiaru pienistości

274 **foam stabiliser**
A product which increases the stability of the foam.
Note: According to the conditions of test or use, or according to the nature of the foaming product, the effect of stabilization can also result in the formation of a greater volume of foam as well as leading to a greater persistence of the foam produced.
fr stabilisateur de mousse
Produit augmentant la stabilité de la mousse.
Nota: Suivant les conditions d'essai ou d'emploi, ou suivant la nature du produit moussant, l'effet de stabilisation peut entraîner la formation d'un plus grand volume de mousse, ainsi que conduire à une plus grande persistance de la mousse formée.
de Schaumstabilisator
Produkt, welches die Stabilität des Schaums erhöht.
Anmerkung: Entsprechend den Versuchsbedingungen kann je nach Art des Schaummittels die stabilisierende Wirkung sowohl in der Bildung einer grösseren Schaummenge als auch in der grösseren Stabilität des bereits vorhandenen Schaumes bestehen.

el estabilizador de espuma
Producto que aumenta la estabilidad de la espuma.
Observación: Según las condiciones de ensayo o de empleo, o según la naturaleza del producto espumante, el efecto de estabilización puede ser causa de la formación de un mayor volumen de espuma, además de conducir a una mayor persistencia de la espuma formada.
it stabilizzatore di schiuma
ne schuimstabilisator
Een stof die de stabiliteit van het schuim verhoogt.
Opmerking: Naar gelang van de omstandigheden bij de proef kan naar de aard van het schuimmiddel het effekt van het stabiliseren zowel een grotere hoeveelheid schuim als ook een grotere stabiliteit van het reeds gevormde schuim zijn.
pl stabilizator piany
Produkt poprawiający trwałość piany.
Uwaga: W zależności od warunków badania lub warunków użycia oraz w zależności od rodzaju środka pieniącego działanie stabilizujące może uwidoczniać się zarówno w tworzeniu się większej objętości piany jak i w większej jej trwałości.

275 **foam stability**
The ability of a foam to persist.
fr stabilité de mousse
Degré d'aptitude d'une mousse à la persistance.
de Schaumstabilität
Grad der Fähigkeit eines Schaumes, beständig zu bleiben.
el estabilidad de espuma
Aptitud de una espuma para la persistencia.
it stabilità di una schiuma
Attitudine di una schiuma alla persistenza.
ne stabiliteit van het schuim
Het vermogen van een schuim zich te handhaven.
pl trwałość piany

276 **foaming**
The action causing formation of foam.
fr moussage
Action entraînant la formation d'une mousse.
de Schäumen
Vorgang, der zur Bildung von Schaum führt.
el espumación
Acción que conduce a la formación de espuma.
it schiumeggiamento
Azione che conduce alla formazione di schiuma.
ne schuimen

De werking die de vorming van schuim veroorzaakt.

pl pienienie, spienianie
Działanie powodujące powstawanie piany.

277 foaming agent, foamer†
A substance which, when introduced into a liquid, confers on it an ability to form foam.

fr agent moussant, moussant
Produit qui, introduit dans un liquide, lui communique une aptitude à la formation de mousse.

de Schaummittel
Produkt, das einer Lösung die Fähigkeit zur Schaumbildung vermittelt.

el producto espumante, espumante
Producto que introducido en un líquido, le comunica una aptitud para formar espuma.

it schiumogeno
Sostanza che sciolta o dispersa in un liquido gli conferisce l'attitudine a formare schiuma.

ne schuimmiddel
Een stof die bij toevoeging een schuimend vermogen verleent.

pl środek pianotwórczy
Substancja, która po wprowadzeniu do cieczy nadaje jej zdolność tworzenia piany.
† See appendix

278 foaming characteristics, lathering properties
fr propriétés moussantes
de Schaumeigenschaften
el propiedades espumantes
it proprietà schiumogena
ne eigenschappen van het schuim
pl własności pianotwórcze

279 foaming power
The ability to produce foam.
fr pouvoir moussant
Degré d'aptitude à former de la mousse.
de Schaumvermögen
Grad der Fähigkeit, Schaum zu bilden.
el poder espumante
Capacidad para formar espuma.
it potere schiumogeno
Grado di attitudine a formare schiuma.
ne schuimvermogen
Het vermogen schuim te vormen.
pl zdolność pianotwórcza, zdolność pienienia się

280 food debris
fr débris d'aliments, débris alimentaires
de Speiserückstände
el residuos de alimentos
it residui di cibo
ne etensresten
pl pozostałości pożywienia, resztki jedzenia

281 foundation make-up
fr fond de teint
de Make-up-Grundlagen
el maquillaje
it fondo tinta
ne onderlaag, grimeersel
pl podkład pod makijaż

282 free adhesion energy (liquid–solid) (symbol A_a)†
The energy which manifests itself in the work required to achieve, in an isothermal and reversible manner, a separation at the interface between two phases (liquid/solid) with the formation of a new free liquid surface of the same dimensions as the initial interface. It is the sum of the free energy of wetting and the free surface energy. It is expressed in Joules (J)[1].

fr énergie libre d'adhésion (liquide–solide) (symbole A_a)†
Énergie qui se manifeste dans le travail à fournir pour provoquer d'une façon isotherme et réversible une séparation à l'interface limitant deux phases liquide-solide avec formation d'une nouvelle surface liquide libre, conservant les mêmes dimensions que l'interface initiale. Elle est la somme de l'énergie libre de mouillage et de l'énergie superficielle libre et s'exprime en joules (J)[1].

de freie Adhäsionsenergie (Symbol A_a)†
Energie, welche in der Arbeit zum Ausdruck kommt, die geleistet werden muss, um isotherm und reversibel eine Trennung an der Grenzfläche zweier Phasen (flüssig/fest) hervorzurufen, wobei eine freie Flüssigkeitsoberfläche mit den Dimensionen der ursprünglichen Grenzfläche gebildet wird. Sie ist die Summe der freien Benetzungsenergie und der freien Oberflächenenergie; sie wird in Joule (J) ausgedrückt[1].

el energía libre de adhesión (líquido–sólido)
Energía que se manifiesta en el trabajo a suministrar para producir de manera

isoterma y reversible una separación en la interfacie entre una fase líquida y otra sólida, con formación de una superficie líquida libre, de las mismas dimensiones que la interfacie inicial. Esta energía es la suma de la energía libre de mojado y de la energía superficial libre.

it energia libera di adesione (liquido–solido) Il lavoro richiesto per ottenere, in condizioni isotermiche, isobariche e reversibili, una separazione all'interfaccia che limita due fasi (liquido-solido) con formazione di una superficie libera liquida avente le stesse dimensioni dell'interfaccia iniziale, si traduce in aumento di energia libera del sistema. Tale energia si chiama energia libera di adesione; corrisponde alla somma dell'energia libera di bagnatura e dell'energia superficiale libera e si esprime in erg.

ne vrije adhesie-energie Energie die tot uitdrukking komt in de arbeid die verricht moet worden om een isotherm en reversibele scheiding aan het grensvlak van twee fasen (vloeibaar/vast) teweeg te brengen, waarbij een vrij vloeistof oppervlak met de afmetingen van het oorspronkelijke grensvlak wordt gevormd. Zij is de som van de vrije bevochtigingsenergie en de vrije oppervlakte energie en wordt in Joules (J) uitgedrukt[1].

pl swobodna energia adhezji (ciecz–ciało stałe) (symbol A_a) Energia objawiająca się koniecznością wykonania pracy dla rozdzielenia w sposób izotermiczny i odwracalny dwóch faz (ciecz/ciało stałe) na granicy faz z utworzeniem nowej swobodnej powierzchni cieczy o wymiarach takich samych jakie miała powierzchnia międzyfazowa. Jest to suma swobodnej energii zwilżania i swobodnej energii powierzchniowej wyrażona w dżulach (1 J = 1 N · 1 m).

[1] 1 J = 10^7 erg.
† See appendix
  Voir appendice
  Siehe Appendix

283 free flowing powder
fr poudre de libre écoulement
de freifliessendes Pulver
el polvo de libre vertido
it polvere scorrevole
ne vrij vloeiend poeder
pl proszek swobodnie spływający, proszek sypki

284 free interfacial energy (liquid–liquid)
The energy which manifests itself in the work required to increase or form the interface separating two liquid phases, in an isothermal and reversible manner. The free interfacial energy is expressed in Joules (J)[1].

fr énergie interfaciale libre (liquide–liquide) Énergie qui se manifeste dans le travail à fournir pour augmenter ou former l'interface séparant deux phases liquides, de façon isotherme et réversible. Elle s'exprime en joules (J)[1].

de freie Grenzflächenenergie (flüssig/flüssig) Energie, welche in der Arbeit zum Ausdruck kommt, die geleistet werden muss, um die Grenzfläche zweier flüssiger Phasen isotherm und reversibel zu vergrössern oder zu bilden. Sie wird in Joule (J) ausgedrückt[1].

el energía interfacial libre Energía que se manifiesta en el trabajo a suministrar para aumentar o formar la interfacie que separa las fases líquidas de manera isoterma y reversible.

it energia interfacciale libera Il lavoro richiesto per aumentare o formare in condizioni isotermiche, isobariche e reversibili l'interfaccia che separa due fasi liquide, si traduce in un aumento di energia libera del sistema. Questo aumento di energia libera si chiama energia interfacciale libera e si esprime in erg.

ne vrije grensvlakenergie Energie die tot uitdrukking komt in de arbeid die nodig is om het grensvlak van twee vloeibare fasen isotherm en reversibel te vergroten of te vormen. Deze wordt uitgedrukt in Joules (J)[1].

pl swobodna energia międzyfazowa (ciecz–ciecz) Energia objawiająca się koniecznością wykonania pracy dla zwiększenia lub utworzenia w sposób izotermiczny i odwracalny powierzchni międzyfazowej rozdzielającej dwie fazy ciekłe. Energię tę wyraża się w dżulach (1J = 1N · 1m).
[1] 1 J = 10^7 erg.

285 free surface energy
The energy which manifests itself in

the work required to increase or form a surface in a liquid, in an isothermal and reversible manner. The free surface energy is expressed in Joules (J)[1].

fr énergie superficielle libre
Énergie qui se manifeste dans le travail à fournir pour augmenter ou former la surface d'un liquide, de façon isotherme et réversible. Elle s'exprime en joules (J)[1].

de freie Oberflächenenergie
Energie, welche in der Arbeit zum Ausdruck kommt, die geleistet werden muss, um die Oberfläche einer Flüssigkeit isotherm und reversibel zu bilden oder zu vergrössern. Sie wird in Joule (J) ausgedrückt[1].

el energía superficial libre
Energía que se manifiesta en el trabajo a suministrar para aumentar o formar la superficie de un líquido de manera isoterma y reversible.

it energia superficiale libera
Il lavoro richieto per aumentare o formare la superficie di un liquido in condizioni isotermiche, isobariche, reversibili, si traduce in un aumento di energia libera del sistema. Questo aumento di energia libera si chiama energia superficiale libera e si esprime in erg[2].

ne vrije oppervlakenergie
Energie die tot uitdrukking komt in arbeid, die verricht moet worden om het oppervlak van een vloeistof isothermisch en omkeerbaar te vormen of te vergroten. Deze energie wordt uitgedrukt in Joules (J)[1].

pl swobodna energia powierzchniowa
Energia objawiająca się koniecznością wykonania pracy dla zwiększenia lub utworzenia powierzchni cieczy w sposób izotermiczny i odwracalny. Energię tę wyraża się w dżulach (1 J = 1 N · 1 m).
[1] 1 J = 10[7] erg.
[2] L'erg è l'unità del sistema CGS. L'unità SI è il Joule (J).

286 free wetting energy†
The energy which manifests itself in the work obtained when a surface is wetted in an isothermal and reversible manner, without changing the size of the free liquid surface. It is expressed in Joules (J) [1].

fr énergie libre de mouillage†

Énergie qui se manifeste dans le travail obtenu lorsqu'on mouille une surface d'une façon isotherme et réversible, sans modification simultanée de la grandeur de la surface liquide libre. Elle s'exprime en joules (J)[1].

de freie Benetzungsenergie†
Energie, welche in der Arbeit zum Ausdruck kommt, die gewonnen wird, wenn eine Oberfläche isotherm und reversibel ohne gleichzeitige Änderung der Flüssigkeitsoberfläche benetzt wird. Sie wird in Joule (J) ausgedrückt[1].

el energía libre de mojado
Energía que se manifiesta en el trabajo obtenido cuando se moja una superficie de manera reversible e isoterma, sin modificación simultánea de la extensión de la superficie líquida libre.

it energia libera di bagnatura
Il lavoro che si ottiene quando si bagna una superficie in condizioni isotermiche, isobariche e reversibili senza modificare la grandezza della superficie liquida libera, si traduce in una diminuzione di energia libera del sistema. Questa diminuzione di energia libera si chiama energia libera di bagnatura e si esprime in erg.

ne vrije energie van het bevochtigen
Energie die in de arbeid tot uitdrukking komt die wordt verkregen wanneer een oppervlak isotherm en reversibel, zonder gelijktijdige verandering van het vloeistofoppervlak wordt bevochtigd. Deze wordt in Joules uitgedrukt[1].

pl swobodna energia zwilżania
Energia objawiająca się pracą uzyskaną przy zwilżaniu powierzchni w sposób izotermiczny i odwracalny, bez jednoczesnej zmiany rozmiaru swobodnej powierzchni cieczy. Energię tę wyraża się w dżulach (1 J = 1 N · 1 m).
[1] 1 J = 10[7] erg.
† See appendix
Voir appendice
Siehe Appendix

287 fulling assistant
Product intended to facilitate the formation of felt during the fulling operation.
Note: Generally, the fibres are made more slippery by surface active agents or preparations containing them, such as: soaps, alkylsulphates and fatty acid condensates, possibly in conjunction with mineral or organic swelling agents.

fr adjuvant de foulage
Produit ayant pour but de faciliter

la formation du feutre pendant l'opération de foulage.

Nota: Il s'agit généralement de faciliter le glissant
des fibres au moyen d'agents de surface ou de
préparations en comportant, tels que: savons,
alkylsulfates et condensats d'acides gras, en
association éventuelle avec des corps gonflants
minéraux ou organiques.

de Walkhilfsmittel
Produkt, das die Aufgabe hat, beim
Walkprozess die Filzbildung zu unter-
stützen.

Anmerkung: Es handelt sich im allgemeinen um
gleitendmachende grenzflächenaktive Stoffe oder
Zubereitungen auf dieser Basis, wie Walkseifen,
Alkylsulfate und Fettsäurekondensationsprodukte mit anorganischen oder organischen quell-
baren Körpern.

el auxiliar de batanado
Producto que tiene por objeto favorecer
el fieltrado durante la operación de
batanado.

Observación: Se trata generalmente de substancias tensioactivas o de preparaciones que las
contienen, que confieren propiedades deslizantes
a las fibras, tales como jabones, alquilsulfatos
y productos de condensación de ácidos grasos,
en posible asociación con substancias hidratables,
minerales u orgánicas.

it —

ne volmiddel
Produkt bedoeld om de vervilting bij
het vollen te bevorderen.

pl środek pomocniczy do spilśniania
Produkt, którego zadaniem jest ułat-
wienie tworzenia się pilśni filcu podczas
operacji spilśniania.

Uwaga: Zwykle włókna stają się bardziej śliskie
dzięki użyciu związków powierzchniowo czynnych
takich jak mydła, alkilosiarczany i produkty
kondensacji kwasów tłuszczowych lub zawierających je preparatów przygotowanych ewentualnie
z dodatkiem nieorganicznych lub organicznych
środków spęczniających.

G

288 general cleaning
fr nettoyage en général
de allgemeine Reinigung
el lavado general
it pulizia in genere
ne wassen in het algemeen
pl ogólne oczyszczanie

289 germicide
fr germicide
de keimtötender Wirkstoff
el germicida
it germicida
ne kiemdodend middel
pl środek bakteriobójczy

290 graining out
The separation into two phases, "curd soap–lye", of the isotropic solution of soap by the action of electrolytes, in order to eliminate excess water and also impurities. In the case of saponification of a neutral fat, this operation also allows the separation of glycerine.

fr relargage
Démixtion en deux phases, "savon grainé–lessive inférieure", de la solution isotrope de savon, sous l'action d'électrolytes, dans le but d'éliminer l'excès d'eau, ainsi que des impuretés. En cas de saponification de corps gras neutres, cette opération permet en plus la séparation de la glycérine.

de Aussalzung
Entmischung des Seifenleims in zwei Phasen: "geronnener Kern–Unterlauge" durch Elektrolytbehandlung, um das überschüssige Wasser sowie Verunreinigungen zu entfernen. Bei der Verseifung von neutralen Fetten erlaubt diese Operation gleichzeitig die Abtrennung des Glycerins.

el saladura
Separación de la disolución isótropa en dos fases, "jabón graneado" y "lejía inferior", bajo la acción de electrólitos, con objeto de eliminar el exceso de agua, así como las impurezas. En el caso de saponificación de grasas, esta operación permite además la separación de la glicerina.

it salatura
La separazione in due fasi "sapone levato–liscivia alcalina" di una soluzione isotropa di sapone a mezzo elettroliti, al fine de eliminare l'eccesso di acqua e le impurezze. Nel caso di saponificazione di grasso neutro, tale operazione porta pure alla separazione della glicerina.

ne uitzouten
Ontmengen van de zeeplijm in twee fasen: "kernzeep–onderloog" door behandeling met electrolyten om de overmaat water alsmede verontreinigingen te verwijderen. Bij het verzepen van neutrale vetten veroorlooft deze operatie tegelijkertijd het afscheiden van de glycerine.

pl wysalanie
Rozdział izotropowego roztworu mydła na dwie fazy, tj. wysół ścięty i ług pomydlany, pod wpływem działania elektrolitu, w celu usunięcia nadmiaru wody i zanieczyszczeń. W przypadku zmydlania tłuszczów obojętnych wysalanie umożliwia również wydzielenie gliceryny.

291 greasy cream
fr crème grasse
de Fettcreme
el crema grasa
it crema grassa
ne vettige crème
pl krem tłusty

292 grey-haired
fr aux cheveux gris, aux poils gris
de grauhaarig
el cabello gris
it capelli grigi
ne grijsharig
pl siwowłosy

293 greying (hair)
fr grisonnement
de Grauwerden
el encanecimiento
it incanutire
ne grijs worden
pl siwienie (włosów)

294 greying (textile)
fr ternissement de la teinte, "grisure"
de Vergrauung
el pardeamiento
it iscurimento del colore

ne	vergrauwen
pl	szarzenie (tkaniny)

295 greying of the hair, blanching of the hair
fr	canitie
de	Ergrauen der Haare
el	encanecer del cabello, blanquear del cabello
it	canizie
ne	vergrijzen van het haar
pl	posiwianie włosów

296 growth of hair
fr	chevelure
de	Haarwuchs
el	cabellera
it	capigliatura
ne	haargroei
pl	wzrost włosa

297 gum disorder
fr	troubles gingivaux
de	Zahnfleischwunde
el	trastornos de las encías
it	malattie delle gengive
ne	tandvleesaandoening
pl	schorzenie dziąseł

H

298 hair
fr cheveu, poil
de Haar
el cabello
it capello, pelo
ne haar
pl włos, włosy

299 hair cream
fr crème à coiffer
de Frisiercreme
el crema para el cabello
it crema per capelli
ne haarcrème
pl krem do włosów

300 hair dye remover
fr décapeur, produit décapant
de Haarfarbenentferner
el desoxidante, producto desoxidante
it prodotto decapante
ne haarverfverwijderer
pl środek do odbarwiania włosów, środek do usuwania farb z włosów

301 hair dyes
fr colorant pour cheveux, colorant capillaire
de Haarfarbstoff
el colorante para el cabello
it tintura per capelli
ne haarverven
pl farby do włosów, barwniki do włosów

302 hair follicle
fr follicule pileux
de Haarfollikel, Haarbalg
el folículo piloso
it follicolo pilifero
ne haarzakje
pl mieszek włosowy

303 hair lacquer
fr laque pour cheveux
de Haarlackprodukt
el laca para el cabello
it lacca per capelli
ne haarlak
pl lakier do włosów

304 hair lotion
fr lotion capillaire
de Haarwasser
el loción capilar
it lozione per capelli

ne haarwater
pl płyn do włosów, woda do włosów

305 hair-setting lotion
fr lotion de mise en plis
de Einlegemittel
el loción para marcar el cabello
it lozione per la messa in piega
ne watergolf
pl płyn do utrwalania fryzury, płyn do układania włosów

306 hair shaft
fr tige capillaire
de Haarschaft
el tronco capilar
it stello del capello
ne haarschacht
pl łodyga włosa, trzon włosa

307 hair straightening cream
fr crème antifrisante, crème pour défriser
de haarglättende Creme
el crema para alisar el cabello
it crema antipieghe
ne sluikmiddel
pl krem do prostowania włosów

308 hair tinting preparation
fr rinçage colorant, colorant de nuançage fugace
de Haartönungsmittel
el preparado para teñir el cabello
it preparato per la tintura dei capelli
ne haartintmiddel
pl preparat do koloryzowania włosów, środek tonujący do włosów

309 hairwaving product
fr produit à permanente, produit pour permanente
de Dauerwellmittel, Haarwell-Produkt
el producto para la permanente
it prodotto per permanente
ne haarkrulmiddel
pl środek do trwałej ondulacji

310 hand creams and lotions
fr crèmes et lotions pour les mains
de Handcremes und flüssige Handcremes
el crema y loción para las manos
it creme e lozioni per le mani
ne handcrèmes en toiletwaters
pl kremy i płyny do rąk

311 hard water
fr eau dure
de Hartwasser, hartes Wasser
el agua dura
it acqua dura
ne hard water
pl woda twarda

312 hard water soap
fr savon pour eaux dures
de Hartwasserseife
el jabón para aguas saladas, jabón para agua de mar, jabón para aguas duras
it sapone per acque dure
ne hardwaterzeep
pl mydło odporne na twardą wodę, mydło dla twardej wody

313 hardness salts, hard water salts
fr sels de dureté de l'eau
de Härtebildner
el sales de dureza del agua
it sali di durezza dell'acqua
ne hardheidvormers
pl sole powodujące twardość wody

314 harmful rays
fr rayons dangereux, rayons nocifs
de bedenkliche Strahlung
el rayos nocivos
it raggi nocivi
ne schadelijke stralen
pl promieniowanie szkodliwe

315 heavy duty detergent
fr détergent pour gros travaux, produit de lavage pour linge gros
de Grobwaschmittel
el detergente para lavados enérgicos, producto de lavado para lino bruto (ropa gruesa)
it detergente per bucato
ne wasmiddel voor de grote was
pl środek do prania o energicznym działaniu

316 heavy laundering
fr gros blanchissage, blanchissage des tissus résistants, blanchissage des vêtements de travail
de Grobwäsche
el colada, lavado de tejidos resistentes, lavado de ropa de trabajo
it grande bucato
ne grote was
pl pranie energiczne

317 home permanent
fr ondulation faite à domicile, permanente à domicile
de Heimdauerwelle
el permanente casera
it ondulazione fatta a casa, permanente a casa
ne blijvende haargolf voor huisgebruik
pl trwała ondulacja domowa

318 horny layer (of the skin)
fr couche cornée (de la peau)
de Hornschicht (der Haut)
el capa córnea de la piel
it strato corneo (della pelle)
ne hoornlaag
pl warstwa rogowa (skóry)

319 household detergent
fr détergent ménager
de Haushaltswaschmittel
el detergente doméstico, detergente casero
it detergente ad uso domestico
ne huishoudwasmiddel
pl środek do prania dla gospodarstwa domowego

320 household performance test, domestic washing test
fr essai de lavage dans la pratique
de Gebrauchswertbestimmung, Gebrauchsprüfung, Praxisversuch
el ensayo de lavado doméstico, ensayo de lavado en la práctica
it prove di lavaggio in pratica
ne wasproef op huishoudschaal
pl praktyczna ocena zdolności piorącej

321 household soap
Soap containing 60–70% fatty acids used for household washing purposes.
fr savon de ménage
Savon employé pour les opérations ménagères de lavage, contenant de 60 à 70% d'acides gras.
de Kernseife, Haushaltseife
Ca. 60–70% Fettsäure enthaltende Seife, die für Wasch- und Reinigungszwecke im Haushalt verwendet wird.
el jabón de lavar
Jabón con 60 a 70% de ácidos grasos, empleado para el lavado y limpieza domésticos.
it sapone per uso domestico
Sapone contenente acidi grassi nella misura del 60–70% destinato ad uso domestico.

ne huishoudzeep
Zeep met 60–70% vetzuur, die voor de huishoudwas dient.
pl mydło gospodarcze
Mydło zawierające 60–70% kwasów tłuszczowych stosowane do prania w gospodarstwach domowych.

322 humectant
fr humidifiant
de Feuchthaltemittel
el humectante
it agente umettante
ne middel tot het op peil houden van het vochtgehalte
pl środek utrzymujący wilgotność skóry

323 hydrogen bond
fr liaison hydrogène
de Wasserstoffbindung
el enlace de hidrógeno
it legame idrogeno
ne waterstofband
pl wiązanie wodorowe

324 hydrogen peroxide
fr eau oxygénée
de Hydrogenperoxid, Wasserstoffperoxid
el agua oxigenada, peróxido de hidrógeno
it acqua ossigenata
ne waterstofperoxide
pl nadtlenek wodoru

325 hydrolysis
Reaction of splitting by water. In the particular case of surface active agents, hydrolysis is more particularly the reverse action to esterification or amide formation and is characterized by the formation of an acid and of an alcohol, enol or phenol or of ammonia or an amine. Hydrolysis of fats gives rise to fatty acids and glycerol; that of soap, to fatty acids and a base.
fr hydrolyse
Réaction de scission par l'eau. Dans le cas particulier des agents de surface, l'hydrolyse est, notamment, la réaction inverse de l'estérification ou de l'amidification et se caractérise alors par la formation d'un acide et d'un alcool, énol ou phénol, ou d'ammoniac ou amine. L'hydrolyse des corps gras conduit aux acides gras et au glycérol, celle des savons aux acides gras et à une base.

de Hydrolyse
Im speziellen Fall grenzflächenaktiver Körper: Aufspaltungsreaktion durch Wasser. Die Hydrolyse ist stets die umgekehrte Reaktion der Veresterung und der Amidierung, charakterisiert durch die Bildung einer Säure und eines Alkohols, Enols oder Phenols, bzw. von Ammoniak oder einem Amin. Die Hydrolyse der Fette führt zur Bildung von Fettsäuren und Glycerin, diejenige der Seifen zu Fettsäuren und einer Base.
el hidrólisis
En el caso particular de los agentes de superficie: reacción de desdoblamiento por el agua. La hidrólisis es la reacción inversa de la esterificación y de la amidificación y se caracteriza por la formación de un ácido y de un alcohol, enol o fenol, o de amoníaco o amina. La hidrólisis de las grasas conduce a los ácidos grasos y a glicerina; la de los jabones, a los ácidos grasos y a una base.
it idrolisi
Nel caso particolare dei tensioattivi: reazione di scissione per mezzo dell'acqua. L'idrolisi è più precisamente la reazione inversa dell'esterificazione o dell'amidazione ed è caratterizzata dalla formazione di un acido e di un alcole (enolo o fenolo), o di ammoniaca o ammina. L'idrolisi dei grassi porta alla formazione di acidi grassi e glicerolo, quella dei saponi a formare acidi grassi e una base.
ne hydrolyse
In het bijzondere geval van oppervlakaktieve stoffen is dit de splitsingsreactie met behulp van water. Het is met name de reactie tegengesteld aan verestering of amidatie en wordt gekenmerkt door de vorming van een zuur en een alcohol, enol of fenol of van ammoniak of een amine. Hydrolyse van vetten geeft vetzuren en glycerol, van zeep vetzuren en een base.
pl hydroliza
Reakcja rozpadu związku chemicznego pod wpływem wody. W przypadku związków powierzchniowo czynnych hydroliza jest reakcją odwrotną do reakcji estryfikacji i amidowania, charakteryzującą się tworzeniem kwasu oraz alkoholu, enolu lub fenolu względnie

amoniaku lub aminy. Hydroliza tłuszczów prowadzi do utworzenia kwasów
tłuszczowych i gliceryny, a hydroliza
mydła do utworzenia kwasów tłuszczowych i zasady.

326 hydrophilic group
Molecular group having an endophilic
behaviour in relation to water.
fr groupement hydrophile
Groupement moléculaire ayant un comportement endophile vis-à-vis de l'eau.
de hydrophile Gruppe
Molekulare Gruppe, die sich gegen
Wasser endophil verhält.
el grupo hidrófilo
Grupo molecular que posee un comportamiento endófilo con respecto al
agua.
it gruppo idrofilo
Gruppo molecolare avente comportamento endofilo rispetto all'acqua.
ne hydrofiele groep
Moleculaire groep met een endofiel
gedrag ten opzichte van water.
pl grupa hydrofilowa
Część cząsteczki wykazująca endofilny
charakter w stosunku do wody. •

327 hydrophilic-lipophilic balance
The relative power of the polar group
or groups and of the non-polar part
conditions the affinities of the molecule
to water and to organic solvents of low
polarity, respectively. The relation between these affinities represents the
hydrophilic-lipophilic ratio of the compound.
Note: This definition relates only to emulsifying
agents.
fr rapport hydro-lipophile
L'importance relative du (ou des)
groupement(s) polaire(s) et de la partie
apolaire conditionne les affinités respectives de la molécule pour l'eau et
pour les solvants organiques peu polaires.
Elle représente le rapport hydro-lipophile du composé.
Nota: Cette définition ne concerne que les
produits émulsionnants (émulsifiants).
de hydro-lipophiles Verhältnis
Das Verhältnis der polaren Gruppe
oder Gruppen und des apolaren Restes
zueinander bedingt die Affinitäten des
Moleküls zum Wasser und zu schwach
polaren organischen Lösungsmitteln. Das
Verhältnis dieser Affinitäten stellt das

hydro-lipophile Verhältnis der Verbindung dar.
Anmerkung: Diese Definition ist nur für Emulga-
toren anwendbar.
el relación hidro-lipófila
Las afinidades respectivas de una molécula para el agua y para los disolventes orgánicos poco polares dependen
de la importancia relativa del grupo
polar o de los grupos polares y del
radical apolar. El cociente de dichas
afinidades representa la relación hidro-
lipófila del compuesto.
Observación: Esta definición sóla se aplica a los
productos emulsionantes.
it rapporto idro-lipofilo
L'importanza relativa del o dei gruppi
polari e della parte determine le rispettive affinità della molecola per
l'acqua e per i solventi organici poco
polari. Il rapporto fra queste affinità
è il rapporto idro-lipofilo del composto.
Nota: Questa definizione riguarda solamente
gli emulsionanti.
ne verhouding tussen de hydrofiele en
lipofiele eigenschappen (HLB-waarde)
De relatieve kracht van de polaire
groep of groepen en van het niet-polaire deel is bepalend voor de affiniteiten van het molecuul voor water,
resp. voor organische stoffen met een
lage polariteit. Het verband tussen
deze affiniteiten is de verhouding tussen
hydrofiele en lipofiele eigenschappen
van de verbinding.
pl równowaga hydrofilno-lipofilna
Względny stosunek grupy lub grup
polarnych i reszty niepolarnej określa
powinowactwo cząsteczki do wody i słabo polarnych rozpuszczalników organicznych. Stosunek tych powinowactw
stanowi stosunek hydrofilno-lipofilny
związku.
Uwaga: Definicja ta dotyczy tylko środków
emulgujących.

328 hydrophily
Endophily in relation to water.
fr hydrophilie
Endophilie vis-à-vis de l'eau.
de Hydrophilie
Endophilie für eine wässrige Phase.
el hidrofilia
Endofilia con respecto al agua.
it idrofilia
Endofilia rispetto all'acqua.

ne hydrofilie
Endofilie ten opzichte van water.
pl hydrofilia
Endofilia w stosunku do wody.

329 hydrophobic group
Molecular group having an exophilic
behaviour in relation to water.
fr groupement hydrophobe
Groupement moléculaire ayant un comportement exophile vis-à-vis de l'eau.
de hydrophobe Gruppe
Molekulare Gruppe, die sich gegen
Wasser exophil verhält.
el grupo hidrófobo
Grupo molecular que posee un comportamiento exófilo con respecto al
agua.
it gruppo idrofobo
Gruppo molecolare avente comportamento esofilo rispetto all'acqua.
ne hydrofobe groep
Moleculaire groep met een exofiel
gedrag t.a.v. water.
pl grupa hydrofobowa
Część cząsteczki o egzofilnym zachowaniu w stosunku do wody.

330 hydrophoby
Exophily in relation to water.
fr hydrophobie
Exophilie vis-à-vis de l'eau.
de Hydrophobie
Exophilie für eine wässrige Phase.
el hidrofobia
Exofilia con respecto al agua.
it idrofobia
Esofilia rispetto all'acqua.
ne hydrofobie
Exofilie ten opzichte van water.
pl hydrofobia
Egzofilia w stosunku do wody.

331 hydrotropy
By hydrotropy is understood the fact
that the solubility of a substance which
is only slightly soluble in water is
increased by the addition of a third
substance. This third substance is
called an "hydrotropic agent".
fr hydrotropie
On entend par hydrotropie le fait
que la solubilité d'une substance peu
soluble dans l'eau est augmentée par
l'addition d'une troisième substance.
Cette troisième substance est appelée
"agent hydrotrope".
de Hydrotropie
Unter Hydrotropie versteht man die
Tatsache, dass die Löslichkeit einer in
Wasser schwerlöslichen Substanz durch
den Zusatz eines dritten Stoffes erhöht
wird. Dieser dritte Stoff wird als "hy-
drotropes Mittel" bezeichnet.
el hidrotropía
Aumento de la solubilidad de una
substancia poco soluble en agua, por
adición de una tercera substancia.
it idrotropia
ne hydrotropie
Men verstaat onder hydrotropie het
feit, dat de oplosbaarheid van een
stof, die weinig oplosbaar in water is,
wordt verhoogd door de toevoeging
van een derde stof. Deze derde stof
wordt "hydrotroop middel" genoemd.
pl hydrotropia
Pod pojęciem hydrotropii rozumie się
zjawisko zwiększenia się rozpuszczalności substancji charakteryzującej się
nieznaczną tylko rozpuszczalnością w
wodzie w wyniku dodania substancji
trzeciej, zwanej "czynnikiem hydro-
tropowym".

I

332 individual susceptibility
fr susceptibilité personnelle
de individuelle Empfindlichkeit
el susceptibilidad personal,
   susceptibilidad individual
it suscettibilità personale
ne individuele gevoeligheid
pl wrażliwość indywidualna, wrażliwość osobnicza

333 industrial soap
Special soap manufactured for a specific industrial application, such as textiles, drawing soaps, etc.
fr savon industriel
   Savon spécial, fabriqué pour des applications techniques spécifiques, comme tréfilerie, textiles, etc.
de Industrieseifen, technische Seifen
   Spezialseifen, die zu besonderen technischen Zwecken verwendet werden, z.B. Drahtziehseifen, Textilseifen, usw.
el jabón industrial
   Jabones especiales empleados para aplicaciones técnicas específicas, como trefilería, industria textil, etc.
it sapone industriale
   Saponi speciali a specifica applicazione industriale (industria tessile, stiratura, ecc.).
ne industrieële zepen
   Speciale zeepsoorten voor technische doeleinden b.v. voor draadtrekken of voor de textielindustrie.
pl mydło techniczne
   Mydło specjalne, wytwarzane dla specyficznych zastosowań przemysłowych takich jak włókiennictwo, ciągnienie drutu itd.

334 initial lather
fr mousse initiale
de Anfangsschaum
el espuma inicial
it schiuma iniziale
ne schuim aan het begin
pl piana początkowa, piana pierwotna

335 interfacial activity
fr activité interfaciale
de Grenzflächenaktivität
el actividad interfacial
it attività interfaciale

ne grensvlakaktiviteit
pl aktywność międzyfazowa

336 interfacial tension (symbols γ and γ_i)†
The force per unit of length, arising from the free interfacial energy. It is expressed in newtons per metre ($N/m = 10^3$ dynes/cm).
fr tension interfaciale (symboles γ et γ_i)†
   Force par unité de longueur, résultante de l'énergie interfaciale libre. Elle s'exprime en newtons par mètre (N/m).
de Grenzflächenspannung (Symbole γ und γ_i)
   Kraft pro Längeneinheit, die aus der freien Grenzflächenenergie resultiert. Sie wird in Newton pro Meter (N/m) ausgedrückt.
el tensión interfacial
   Fuerza por unidad de longitud, resultante de la energía interfacial libre. Numéricamente es igual a la energía interfacial por unidad de superficie.
it tensione interfacciale
   Forza per unità di lunghezza, risultante dall'energia interfacciale libera. E' numericamente uguale all'energia interfacciale libera per unità superficiale dell'interfaccia e si esprime in dine per centimetro.
ne grensvlakspanning
   De kracht per lengte-eenheid voortkomende uit de vrije grensvlakenergie. Deze wordt uitgedrukt in Newtons per meter (N/m).
pl napięcie międzyfazowe (symbole γ i γ_j)
   Siła na jednostkę długości, wynikająca ze swobodnej energii międzyfazowej. Wyraża się ją w niutonach na metr (N/m).
   † See appendix
   Voir appendice

337 invariant zone
The region of a phase diagram which corresponds to the coexistence of three phases. In soap diagrams, the invariant zone of triangular shape, corresponds to the following main equilibria:
— finished soap–middle soap–nigre
— finished soap–nigre–lye
— finished soap–curd soap–lye
The different points within an invariant zone correspond to equilibria of three phases of constant composition where

only the proportion of the phases varies.

fr zone d'invariance
Région d'un diagramme de phases, qui correspond à la coexistence de trois phases. Dans les diagrammes du savon: domaine invariant de forme triangulaire, correspondant aux équilibres principaux suivants:
— savon lisse–savon médian–gras
— savon lisse–gras–lessive inférieure
— savon lisse–savon grainé–lessive inférieure
Les différents points à l'intérieur d'une zone d'invariance correspondent à des équilibres de trois phases à composition constante, où seule la proportion des phases est variable.

de invariante Zone
Gebiet eines Phasendiagrammes, in welchem drei Phasen im Gleichgewicht stehen. Im Seifendiagramm: invariante Bereiche folgender Gleichgewichte:
— geschliffener Seifenkern–Klumpseife–Leimniederschlag
— geschliffener Seifenkern–Leimniederschlag–Unterlauge
— geschliffener Seifenkern–geronnener Seifenkern–Unterlauge
Den Punkten im Innern einer invarianten Zone entsprechen Dreiphasengleichgewichte konstanter Zusammensetzung, bei denen nur das Mengenverhältnis der Phasen veränderlich ist.

el punto triple
Punto de un diagrama de fases en el cual, bajo determinadas condiciones, tres fases pueden encontrarse en equilibrio. En los diagramas del jabón, los puntos triples vienen determinados por la coexistencia de las fases siguientes en equilibrio:
— jabón liquidado–jabón medio–bajos
— jabón liquidado–bajos–sublejía
— jabón liquidado–jabón graneado–sublejía

it zona tre fasi
La regione nel diagramma delle fasi che corrisponde alla coesistenza di tre fasi. Nel diagramma del sapone la zona delle tre fasi è di modello triangolare e corrisponde ai seguenti tre equilibri principali:
— sapone finito–sapone mediano–colletta
— sapone finito–colletta–liscivia alcalina
— sapone finito–sapone levato–liscivia
I differenti punti entro la zona a tre fasi, corrispondono ai vari equilibri delle tre fasi di composizione costante, dove variano solo i rapporti delle fasi.

ne invariante zone
Gebied van een fase-diagram, waarin drie fasen met elkaar in evenwicht zijn. In zeepdiagrammen: in variante gebieden van de volgende evenwichten:
— afgemaakte kernzeep–bonkzeep–lijmzeep
— afgemaakte kernzeep–lijmzeep–onderloog
— afgemaakte kernzeep–vaste kernzeep–onderloog

pl obszar trójfazowy
Obszar wykresu fazowego mydła odpowiadający współistnieniu trzech faz. Na wykresach fazowych mydła obszar trójfazowy o trójkątnym kształcie odpowiada następującym głównym stanom równowagowym:
— wysół płynny–mydło zgęstniałe–klej mydlany
— wysół płynny–klej mydlany–ług spodni
— wysół płynny–wysół ścięty–ług spodni
Punkty wewnątrz obszaru trójfazowego odpowiadają stanom równowagowym trzech faz o stałym składzie, z tym że jedynie proporcje tych faz ulegają zmianie.

338 isoelectric point
fr point isoélectrique
de isoelektrischer Punkt
el punto isoeléctrico
it punto isoelettrico
ne iso-electrisch punt
pl punkt izoelektryczny

J K

339 **judge the "feel"**
fr appréciation du "toucher"
de Griffbeurteilung
el apreciación del "tacto", juzgar del "tacto"
it valutazione del "tatto"
ne beoordeling van de "greep"
pl ocena dotyku, ocena chwytu

340 **keeping properties**
fr propriétés de stockage
de Lagerungseigenschaften
el propiedades de almacenaje
it proprietà di stoccaggio
ne eigenschappen bij opslag
pl przydatność do magazynowania

341 **keratin**
fr kératine
de Keratin
el queratina
it cheratina
ne keratine
pl keratyna

342 **kier boiling assistant**
Product intended to increase the effectiveness and speed of the processing, under pressure or otherwise, of textile materials or articles made of natural or regenerated cellulose fibres, either alone or mixed, with alkaline lye, water, salt solutions or acid solutions. It is applied, for example, in the boiling of grey cotton (under pressure or otherwise), the scouring of linen, the rendering of cotton articles absorbent to water by the continuous process, etc.
Note: These are generally special wetting products, often mixed with solvents.

fr adjuvant de débouillissage et d'hydrophilisation
Produit destiné à rendre plus efficace et plus rapide le traitement, avec ou sans pression, des matières ou articles textiles en fibres cellulosiques naturelles ou régénérées, seules ou en mélange entre elles ou avec des fibres synthétiques, par des lessives alcalines, de l'eau, des solutions salines ou des solutions acides. Il trouve son application par exemple dans le débouillissage du coton écru (avec ou sans pression), le lessivage du lin, l'hydrophilisation des articles coton par procédé à la continue, etc.
Nota: Il s'agit en général de produits mouillants spéciaux souvent en mélange avec des solvants.

de Abkochhilfsmittel und Mittel zum Hydrophilieren
Produkt, das dazu bestimmt ist, die Vorbehandlung von Textilmaterialien aus natürlichen oder regenerierten Cellulosefasern für sich allein oder gemischt mit

Chemiefasern — ohne oder mit Druck — mit Hilfe alkalischer Bäder, salzhaltiger oder saurer Lösungen wirksamer und schneller zu gestalten. Sie werden zum Beispiel zum Abkochen von Rohbaumwolle (ohne oder mit Druck), bei der Vorreinigung von Leinen oder beim Hydrophilieren von Baumwollwaren im Kontinueverfahren angewandt.

Anmerkung: Es handelt sich in der Regel um spezielle Netzmittel, die oft in Mischung mit Lösungsmitteln vorliegen.

el auxiliar para el descrudado y la hidrofilización
Producto destinado para hacer más eficaz y más rápido el tratamiento, con o sin presión, por medio de lejías alcalinas, y de disoluciones salinas o ácidas, de materias textiles constituidas por fibras celulósicas, naturales o regeneradas, solas o mezcladas con fibras sintéticas. Se aplica, por ejemplo, en el descrudado del algodón en rama (con o sin presión), en el lejiado del lino o en la hidrofilización de los artículos de algodón por e procedimiento a la continua.

Observación: Se trata en general de productos humectantes especiales, frecuentemente en mezclas con disolventes.

it —

ne afkookhulpmiddel
Produkt dat tot doel heeft de effektiviteit en snelheid van het afkoken al of niet onder druk te vergroten.

pl środek pomocniczy do warzenia materiałów tekstylnych przed barwieniem
Produkt, którego zadaniem jest zwiększenie efektywności i szybkości ciśnieniowej lub bezciśnieniowej obróbki wyrobów włókienniczych otrzymanych z naturalnej lub regenerowanej celulozy lub jej mieszanek z włóknami chemicznymi, w kąpielach alkalicznych, wodzie oraz roztworach soli lub kwasów. Jest on stosowany np. do ciśnieniowego lub bezciśnieniowego warzenia surowej bawełny, przy oczyszczaniu wstępnym lnu oraz do nadania higroskopijności wyrobom bawełnianym w procesie ciągłym.

Uwaga: Zwykle są to środki zwilżające specjalnego przeznaczenia, często zmieszane z rozpuszczalnikami.

343 kind of soil, type of soil
fr nature de la salissure, type de l'encrassement
de Schmutzart, Art der Anschmutzung

el clase de suciedad, tipo de suciedad
it natura dello sporco, tipo di sporco
ne aard van het vuil
pl rodzaj brudu, typ zabrudzenia

344 kissproof
fr permet le baiser, à l'épreuve du baiser
de kussecht
el no mancha al besar, indeleble
it indelebile
ne bestand tegen kussen
pl odporny na pocałunek

345 Krafft point (symbol t_K and T_K)†
The temperature (in practice, narrow range of temperature) at which the solubility of a ionic surface active agent reaches the critical value for the formation of micelles. Commencing at this temperature, the solubility curve shows a rapid increase.

Note: In the soap industry, Krafft point is the temperature at which a transparent soap solution becomes cloudy on cooling.

fr point de Krafft (symbole t_K et T_K)
Température (pratiquement étroit intervalle de température) à laquelle la solubilité des agents de surface ioniques atteint la valeur de la concentration critique pour la formation de micelles. A partir de cette température, la courbe de solubilité augmente brusquement.

Nota: Dans l'industrie des savons, on désigne par point de Krafft la température à laquelle une solution transparente de savon devient trouble par refroidissement.

de Krafft-Punkt (Symbole t_K und T_K)
Temperatur (praktisch enger Temperaturbereich), bei welcher die Löslichkeit ionischer Tenside den Wert der kritischen Micellbildungskonzentration erreicht. Oberhalb dieser Temperatur steigt die Löslichkeitskurve stark an.

Anmerkung: In der Seifenindustrie bezeichnet man als Krafft-Punkt die Temperatur, bei welcher sich eine transparente Seifenlösung während des Abkühlungsvorganges trübt.

el punto de Krafft
Temperatura (o más exactamente, pequeño intervalo de temperatura) a la cual la solubilidad de los agentes de superficie iónicos alcanza el valor de la concentración crítica para la formación de micelas. A partir de esta temperatura la curva de solubilidad aumenta bruscamente.

Observación: En la industria jabonera se denomina punto de Krafft la temperatura a la cual una disolución transparente de jabón se enturbia por enfriamiento.

it punto di Krafft
Temperatura (più esattamente lo stretto intervallo di temperatura) alla quale una soluzione trasparente di sapone o di certi tensioattivi ionici, diviene torbida per raffreddamento.

Nota: 1. Entro limiti piuttosto ampi questa temperatura è indipendente dalla concentrazione del tensioattivo.
2. Solo nel caso dei saponi di sodio, questa temperatura è vicina è leggermente inferiore alla temperatura di fusione degli acidi grassi impiegati nella loro preparazione.

ne Kraffpunt
Temperatuur (in de praktijk een kort temperatuurtraject) waarbij de oplossing van een oppervlakaktieve stof de kritische micelconcentratie bereikt. Boven deze temperatuur neemt de oplosbaarheid snel toe.

Noot: In de zeepindustrie is het Krafftpunt de temperatuur waarbij een heldere zeepoplossing bij afkoelen ondoorzichtig wordt.

pl punkt Kraffta (symbol t_K i T_K)
Temperatura (w praktyce wąski przedział temperatur), w której rozpuszczalność jonowego związku powierzchniowo czynnego osiąga wartość krytyczną dla tworzenia się miceli. Począwszy od tej temperatury krzywa rozpuszczalności wykazuje gwałtowny wzrost.

Uwaga: W przemyśle mydlarskim punkt Kraffta jest to temperatura, w której przeźroczyste mydło przy schładzaniu ulega zmętnieniu.

† See appendix

L

346 laboratory washing machine
fr machine à laver de laboratoire
de Laboratoriumswaschmachine
el máquina de lavar del laboratorio, lavadora de laboratorio
it macchina lavatrice da laboratorio
ne laboratorium-wasmachine
pl pralnica laboratoryjna, laboratoryjna maszyna do prania

347 laboratory washing test
fr essai de lavage à l'échelle laboratoire, essai de lavage „in labo"
de Laborwaschversuch
el ensayo de lavado en el laboratorio
it prove di lavaggio in laboratorio
ne laboratorium-wasproef
pl laboratoryjny test pralniczy

348 lather behaviour
fr tenue de mousse
de Schaumverhalten
el comportamiento de la espuma
it durata della schiuma
ne gedrag van het schuim
pl zachowanie się piany

349 lather booster
fr renforçateur de mousse, exalteur de mousse
de Schaumverbesserer
el mejorador de espuma, aumentador de espuma, favorecedor de espuma
it esaltatore di schiuma
ne schuimverbeteraar
pl środek wspomagający pienienie, wspomagacz pienienia

350 lather collapse, foam breakage
fr chute de la mousse, brisage de la mousse
de Zusammenbruch des Schaumes
el rompimiento de la espuma
it caduta della schiuma
ne breken van het schuim
pl przełamanie piany

351 lather value
fr indice de mousse
de Schaumwert, Schaumzahl
el índice espumante
it indice di schiuma
ne schuimgetal
pl liczba pianowa

352 lather value in presence of dirt
fr indice de mousse chargée de salissure

de Belastungsschaumzahl
el índice espumante en presencia de suciedad
it indice di schiuma in presenza di sporco
ne belast schuimgetal
pl liczba pianowa w obecności brudu

353 levelling agent
Product designed to promote the even dyeing of textiles.

Note: These products are surface active agents or preparations comprising them, such as: sulphated oils, esters and amides of fatty acids, fatty acid condensates, alkylsulphates, alkylarylsulphonates, alkyl and alkylaryl polyglycolethers, polyglycol esters of fatty acids, amine derivatives. Agents with the properties of protective colloids, such as fatty acid and protein condensates, can also be used.

fr agent égalisant
Produit destiné à favoriser l'unisson tinctorial des textiles.

Nota: Il s'agit d'agents de surface ou de préparations en comportant, tels que: huiles sulfatées, esters et amides d'acides gras, condensats d'acides gras, alkylsulfates, alkylarylsulfonates, alkyl- et alkylarylpolyglycoléthers, esters de polyglycols d'acides gras, dérivés d'amines. Des agents à propriétés protectrices colloïdales, tels que des condensats d'acides gras et de protéines, peuvent également être utilisés.

de Egalisiermittel
Produkt, das dazu bestimmt ist, die gleichmässige Anfärbung des Textilgutes zu fördern.

Anmerkung: Es handelt sich um grenzflächenaktive Stoffe oder Zubereitungen hieraus, wie sulfierte Öle, Fettsäureester und Fettsäureamide, Fettsäurekondensationsprodukte, Alkylsulfate, Alkylarylsulfonate, Alkyl- und Alkylarylpolyglykoläther und Fettsäurepolyglykolester sowie Aminderivate. Es können auch Mittel mit schutzkolloiden Eigenschaften, wie z.B. Fettsäure-Eiweisskondensationsprodukte, verwendet werden.

el agente igualador
Producto destinado a favorecer la tintura uniforme de los textiles.

Observación: Se trata de agentes de superficie, o de preparaciones que los contienen. Se utilizan aceites sulfatados, ésteres y amidas de ácidos grasos, productos de condensación de ácidos grasos, alquilsulfatos, alquilarilsulfonatos, éteres de alquil y de alquilarilpoliglicoles, poliglicolésteres de ácidos grasos, así como derivados de aminas. También pueden utilizarse agentes con propiedades de protección coloidal, tales como productos de condensación de ácidos grasos con proteínas.

it —
ne egaliseermiddel
Een produkt dat het gelijkmatig verven van textiel bevordert.

pl środek wyrównujący, środek egalizujący

Produkt zapewniający równomierne wybarwienie wyrobów włókienniczych.

Uwaga: Produktami takimi są związki powierzchniowo czynne lub zawierające je preparaty. Spośród związków powierzchniowo czynnych w rachubę wchodzą siarczanowane oleje, estry i amidy kwasów tłuszczowych, produkty kondensacji kwasów tłuszczowych, alkilosiarczany, alkiloarylosulfoniany, etery alkilowe i alkiloarylowe poliglikoli, estry poliglikoli i kwasów tłuszczowych oraz pochodne amin. Stosowane mogą być też środki o własnościach koloidów ochronnych takie jak produkty kondensacji kwasów tłuszczowych z białkami.

354 light duty detergent
- **fr** détergent pour articles textiles délicats, produit au lessivage fin, produit pour lavages délicats, produit de lavage léger, produit de lavage pour petit linge
- **de** Feinwaschmittel
- **el** detergente para artículos textiles delicados, producto para lavados finos, producto para lavados delicados, producto para lavados ligeros
- **it** detergente per tessuti delicati
- **ne** wasmiddel voor de fijne was
- **pl** środek do prania tkanin delikatnych, środek do prania tkanin wrażliwych

355 lighten the skin
- **fr** éclaircir la peau
- **de** Haut aufhellen
- **el** aclarar la piel
- **it** schiarire la pelle
- **ne** huid lichter maken
- **pl** wybielać skórę

356 lightproof, fast to light
- **fr** solide à la lumière, inaltérable à la lumière
- **de** lichtecht
- **el** sólido a la luz, fijo a la luz, resistente a la luz
- **it** solidità alla luce
- **ne** lichtbestendig
- **pl** odporny na światło

357 lime soap
- **fr** savon de chaux
- **de** Kalkseife
- **el** jabón de cal, jabón cálcico, jabón calcáreo
- **it** sapone di calcio, sale calcareo
- **ne** kalkzeep
- **pl** mydło wapniowe

358 lime soap dispersion
- **fr** dispersion des savons de chaux
- **de** Kalkseifendispergierung
- **el** dispersión del jabón cálcico
- **it** dispersione dei saponi di calcio
- **ne** dispergeren van de kalkzeep
- **pl** dyspergowanie mydeł wapniowych

359 limiting lye
Lye having a concentration at which soap starts to dissolve. The concentration of the limiting lye depends on the nature of the fatty matter saponified and that of the electrolyte and is a characteristic of the fatty material used in soap manufacture.
- **fr** lessive limite de solubilité du savon (lessive limite)
Lessive possédant une concentration à laquelle le savon commence à se dissoudre. La concentration de la lessive limite dépend de la nature du corps gras saponifié et de celle de l'électrolyte, et est une caractéristique des corps gras utilisés en savonnerie.
- **de** Grenzlauge
Elektrolytlösung einer Konzentration, oberhalb derer die Seife unlöslich ist, unterhalb derer sie sich zu lösen beginnt. Ihre Konzentration hängt von der Natur der verwendeten Fettstoffe, der Elektrolyte und von der Temperatur ab.
- **el** lejía límite de solubilidad del jabón (lejía límite)
Disolución de electrólitos a una concentración por encima de la cual el jabón es insoluble y por debajo de la cual comienza a disolverse. Esta concentración de equilibrio depende de la naturaleza de la grasa empleada, del electrólito y de la temperatura.
- **it** liscivia limite
Una liscivia avente una concentrazione alla quale il sapone incomincia a sciogliersi. La concentrazione della liscivia limite dipende dalla natura degli acidi grassi saponificati, e da quella dell'elettrolita ed è una caratteristica della composizione degli acidi grassi usati nella preparazione del sapone.
- **ne** grensloog
Loog van een concentratie, waarboven de zeep onoplosbaar is en onder de welke hij weer begint op te lossen. Deze concentratie hangt van de aard der gebruikte vetten en de electrolyt af alsmede van de temperatuur.
- **pl** ług o stężeniu granicznym
Roztwór elektrolitu o stężeniu, przy którym rozpoczyna się rozpuszczanie

mydła. Stężenie to zależne od rodzaju zmydlanej substancji tłuszczowej i od rodzaju elektrolitu jest cechą charakterystyczną substancji tłuszczowej stosowanej do produkcji mydła.

360 lipophilic group
Molecular group having an endophilic behaviour in relation to a non-gaseous organic phase.
fr groupement lipophile
Groupement moléculaire ayant un comportement endophile vis-à-vis d'une phase organique non gazeuse.
de lipophile Gruppe
Molekulare Gruppe, die sich gegen eine organische, nichtgasförmige Phase endophil verhält.
el grupo lipófilo
Grupo molecular que posee un comportamiento endófilo con respecto a una fase orgánica no gaseosa.
it gruppo lipofilo
Gruppo molecolare avente comportamento endofilo rispetto ad una fase organica non gassosa.
ne lipofiele groep
Moleculaire groep met een endofiel gedrag ten opzichte van een niet-gasvormige organische fase.
pl grupa lipofilowa
Część cząsteczki o endofilnym zachowaniu w stosunku do niegazowej fazy organicznej.

361 lipophily
Endophily in relation to a non-gaseous non-polar organic phase.
fr lipophilie
Endophilie vis-à-vis d'une phase organique non gazeuse apolaire.
de Lipophilie
Endophilie für eine organische, unpolare, nichtgasförmige Phase.
el lipofilia
Endofilia con respecto a una fase orgánica no gaseosa apolar.
it lipofilia
Endofilia rispetto ad una fase organica non gassosa apolare.
ne lipofilie
Endofilie ten opzichte van een organische, niet-polaire, niet-gasvormige fase.
pl lipofilia
Endofilia w stosunku do niegazowej, niepolarnej fazy organicznej.

362 lipophoby
Exophily in relation to a non-gaseous non-polar organic phase.
fr lipophobie
Exophilie vis-à-vis d'une phase organique non gazeuse apolaire.
de Lipophobie
Exophobie für eine organische, unpolare, nichtgasförmige Phase.
el lipofobia
Exofilia con respecto a una fase orgánica no gaseosa apolar.
it lipofobia
Esofilia rispetto ad una fase organica non gassosa apolare.
ne lipofobie
Exofilie ten opzichte van een organische, niet-polaire, niet-gasvormige fase.
pl lipofobia
Egzofilia w stosunku do niegazowej, niepolarnej fazy organicznej.

363 lipstick
fr rouge à lèvres
de Lippenstift
el lápiz de labios
it rossetto per labbra
ne lippenstift
pl kredka do ust, pomadka do ust

364 lipstick brush
fr pinceau pour rouge à lèvres
de Lippenstift-Pinsel
el pincel para el rojo de labios
it pennellino per rossetto da labbra
ne lippenstiftpenseel
pl pędzelek do warg

365 liquid cream
fr crème liquide
de flüssige Creme
el crema líquida
it crema liquida
ne vloeibare crème
pl krem płynny

366 liquid dentifrice
fr dentifrice liquide, eau dentifrice
de flüssiges Zahnputzmittel
el dentífrico líquido, agua dentífrica
it dentifricio liquido, acqua dentifricia
ne vloeibaar tandverzorgingsmiddel
pl płynny środek do czyszczenia zębów

367 liquid detergent
fr détergent liquide
de flüssiges Waschmittel

el detergente líquido
it detergente liquido
ne vloeibaar wasmiddel
pl środek do prania w płynie, płynny środek do prania, płyn do prania

368 liquid shaving soap
fr savon à barbe liquide
de flüssige Rasierseife
el jabón de afeitar líquido
it sapone da barba liquido
ne vloeibare scheerzeep
pl płynne mydło do golenia

369 liquid soap
fr savon liquide
de flüssige Seife
el jabón líquido
Jabón potásico, obtenido generalmente con aceite de coco o de palmiste, con un contenido de ácidos grasos de 15 a 25%.
it sapone liquido
ne vloeibare zeep
pl mydło płynne, mydło w płynie

370 loading of fabrics
fr incrustation des tissus
de Gewebeinkrustierung
el incrustación en los tejidos
it incrostazione dei tessuti
ne incrustatie der weefsels
pl inkrustacja tkaniny, obciążenie tkaniny

371 loss in tensile strength
fr diminution de la force
de Festigkeitsverlust
el disminución de la fuerza, pérdida de la fuerza
it diminuizione della resistenza meccanica del tessuto
ne achteruitgang in sterkte
pl strata wytrzymałości na rozciąganie

372 loss of hair, alopecia
fr chute des cheveux, chute du poil
de Haarausfall, Haarschwund
el caída del cabello, alopecia
it caduta del pelo, caduta dei capelli
ne haaruitval
pl wypadanie włosów, łysienie

373 low foaming detergent, controlled sudsing detergent, restricted sudsing detergent, low sudser, low foamer
fr détergent peu moussant, détergent à faible pouvoir moussant
de wenig schäumendes Waschmittel, be-grenzt schäumendes Waschmittel, Waschmittel mit kontrollierter Schaummenge
el detergente poco espumante
it detergente poco schiumogeno, detergente a basso potere schiumogeno
ne weinig schuimend wasmiddel
pl słabo pieniący środek piorący

374 lustre
fr lustre
de Glanz
el brillo
it lucentezza
ne glans
pl połysk

375 lustre measurement
fr mesure du lustre
de Glanzmessung
el medida del brillo
it misura della brillantezza
ne glansmeting
pl pomiar połysku

376 lustre production
fr lustrage
de Glänzendmachen
el abrillantamiento
it produzione di brillantezza
ne verwekken van glans
pl nadawanie połysku, nabłyszczanie

377 lye
A solution of electrolytes, practically soap free, separated from the curd soap by graining out and washing.
fr lessive inférieure
Solution d'électrolytes, ne contenant pratiquement pas de savon, séparée du savon grainé par relargage et par lavage.
de Unterlauge
Die beim Aussalzen und Waschen anfallende Elektrolytlösung, welche praktisch seifenfrei ist und Lauge, Verunreinigungen und eventuel Glycerin enthält.
el sublejía
Disolución de electrólitos separada durante la saladura, que contiene débiles cantidades de jabón (0,5 a 1,0%), de álcali y de glicerina (si se ha partido de grasas neutras), así como de impurezas (por ejemplo: restos de tejidos celulares).
it lisciccia inferiore
Una soluzione di elettroliti praticamente esente da sapone e separata dal sapone levato a mezzo salatura e lavaggio.

ne onderloog
De na het uitzouten en wassen als
bijprodukt verkregen electrolyt-oplossing,
die praktisch zeepvrij is en loog, verontreinigingen en eventueel glycerine bevat.
pl ług spodni, ług pomydlany
Roztwór elektrolitów, praktycznie wolny
od mydła, wydzielony z wysołu ściętego
przy wysalaniu i przemywaniu.

378 lyophilic group
A molecular group which has an endophilic behaviour in relation to a liquid
phase.
fr groupement lyophile
Groupement moléculaire ayant un comportement endophile vis-à-vis d'une
phase liquide.
de lyophile Gruppe
Molekulare Gruppe, welche sich gegenüber einer flüssigen Phase endophil
verhält.
el grupo liófilo
it —
ne lyofiele groep
Moleculaire groep met een endofiel
gedrag ten opzichte van een vloeibare
fase.
pl grupa liofilowa
Część cząsteczki o endofilnym zachowaniu w stosunku do fazy ciekłej.

379 lyophily
A predominant tendency to endophily
on the part of matter dispersed in a
medium.
fr lyophilie
Tendance prédominante à l'endophilie
d'une matière répartie dans un milieu.
de Lyophilie
Vorherrschende Tendenz zur Endophilie
einer in einem flüssigen Medium verteilten
Substanz.
el liofilia
Tendencia predominante a la endofilia
de una substancia repartida en un medio.
it liofilia
ne lyofilie
Overheersende neiging tot endofilie van
een in een medium verdeelde stof.
pl liofilia
Dominująca skłonność substancji roz-
proszonej w środowisku ciekłym do
endofilii.

380 lyophobic group
A molecular group which has an exophilic
behaviour in relation to a liquid phase.
fr groupement lyophobe
Groupement moléculaire ayant un comportement exophile vis-à-vis d'une phase
liquide.
de lyophobe Gruppe
Molekulare Gruppe, welche sich gegenüber einer flüssigen Phase exophil verhält.
el grupo liófobo
it —
ne lyofobe groep
Moleculaire groep met een exofiel gedrag
t.a.v. een vloeibare fase.
pl grupa liofobowa
Część cząsteczki o egzofilnym zachowaniu w stosunku do fazy ciekłej.

381 lyophoby
A predominant tendency to exophily on
the part of matter dispersed in a medium.
fr lyophobie
Tendance prédominante à l'exophilie
d'une matière répartie dans un milieu.
de Lyophobie
Vorherrschende Tendenz zur Exophilie
einer in einem flüssigen Medium verteilten Substanz.
el liofobia
Tendencia predominante a la exofilia
de una substancia repartida en un medio.
it liofobia
ne lyofobie
Overheersende neiging tot exofilie van
een in een medium verdeelde stof.
pl liofobia
Dominująca skłonność substancji rozpro-
szonej w środowisku ciekłym do egzofilii.

382 lyotropy
By lyotropy is understood the fact that
the solubility of a substance which is
only slightly soluble in a solvent is
increased by the addition of a third
substance. This third substance is called
a "lyotropic agent".
fr lyotropie
On entend par lyotropie le fait que la
solubilité d'une substance peu soluble
dans un solvant est augmentée par
l'addition d'une troisième substance
Cette troisième substance est appelée
"agent lyotrope".

de Lyotropie
Unter Lyotropie versteht man die Tatsache, dass die Löslichkeit einer in einem Lösungsmittel schwerlöslichen Substanz durch die Zugabe eines dritten Stoffes erhöht wird. Dieser dritte Stoff wird als "lyotropes Mittel" bezeichnet.

el liotropía
Aumento de la solubilidad de una substancia poco soluble en un disolvente, por adición de una tercera substancia.

it liotropia

ne lyotropie
Men verstaat onder lyotropie het feit dat de oplosbaarheid van een stof, die weinig oplosbaar is in een oplosmiddel, verhoogd wordt door de toevoeging van een derde stof. Deze derde stof wordt "lyotroop middel" genoemd.

pl liotropia
Podwyższanie rozpuszczalności substancji charakteryzującej się niewielką tylko rozpuszczalnością przez dodanie substancji trzeciej zwanej "środkiem liotropowym".

M

383 machine washing
fr lavage à la machine
de Maschinenwäsche
el lavado a máquina
it lavaggio a macchina
ne machinewas
pl pranie maszynowe

384 make-up
fr maquillage
de Make-up
el maquillaje
it maquillage
ne opmaaksel
pl makijaż, środek do makijażu

385 make-up for the eyelashes
fr maquillage pour les cils
de Make-up für die Wimpern
el maquillaje para las pestañas
it maquillage per le ciglia
ne opmaaksel voor de oogwimpers
pl kosmetyki do rzęs

386 mascara
fr mascara
de Maskara
el máscara
it mascara
ne mascara
pl tusz do rzęs i brwi

387 mechanical action
fr agitation mécanique, action mécanique
de mechanische Einwirkung
el acción mecánica
it agitazione meccanica, azione meccanica
ne mechanische beweging
pl działanie mechaniczne

388 medicated shampoo
fr shampooing médicamenteux
de medizinisches Shampoo
el champu medicinal
it shampoo medicamentoso
ne medicinal shampoo
pl szampon leczniczy

389 medicated soap
Soap containing pharmaceutical products imparting a therapeutic effect on the skin.
fr savon médicinal
Savon contenant des produits pharmaceutiques produisant un effet thérapeutique sur la peau.

de medizinische Seife
Seife, die durch medikamentöse Zusätze eine therapeutische Wirkung auf die Haut ausübt.
el jabón medicinal
Jabón que, gracias a la adición de medicamentos, posee una acción terapéutica sobre la piel.
it sapone medicinale
Saponi contenenti prodotti farmaceutici con effetto terapeutico sulla pelle.
ne medicinale zeep
Medicament bevattende zeep, die een therapeutische werking op de huid uitoefent.
pl mydło lecznicze
Mydło zawierające produkty farmaceutyczne działające leczniczo na skórę.

390 medulla
fr moelle
de Markzelle, Medulla
el médula
it midollo
ne merg
pl rdzeń, szpik

391 mercerizing assistant
Product used to improve the wetting power of mercerizing lyes and thus to speed up their uniform penetration into the fibres.
Note: These products are wetting agents which are stable in highly concentrated lyes; they are based both on a component which is effective as a surface active agent and as an emulsifier in lyes (alkylsulphates of low molecular weight, highly sulphated oils, cresols, xylenols) and an anti-foaming and wetting substance, made soluble by hydrotropy (for example butyl glycol, ethoxylated amines, etc.).
fr adjuvant de mercerisage
Produit servant à améliorer le pouvoir mouillant des lessives de mercerisage et à accélérer ainsi leur pénétration uniforme dans les fibres.
Nota: Il s'agit d'agents mouillants stables dans des lessives fortement concentrées; ils sont à base, d'une part, d'un composant ayant une efficacité tensioactive et émulsionnante dans les lessives (alkylsulfates à basse masse moléculaire, huiles hautement sulfatées, crésols, xylénols), et d'autre part, d'une substance antimoussante et mouillante, solubilisée par hydrotropie (par exemple: butylglycol, amines éthoxylées, etc.).
de Mercerisierhilfsmittel
Produkt, das dazu dient, die Netzfähigkeit der Mercerisierlauge zu verbessern und

dadurch ihr gleichmässiges Eindringen in die Faser zu beschleunigen.

Anmerkung: Es handelt sich um in hochkon-zentrierten Laugen beständige Netzmittel, die meist eine in der Lauge oberflächenaktive und emulgierend wirkende Komponente (niedrig-molekulare Alkylsulfate, hochsulfierte Öle, Kre-sole, Xylenole) und eine entschäumend und netzend wirkende, hydrotrop gelöste Substanz (z.B. Butylglykol, oxäthylierte Amine u. dgl.) enthalten.

el auxiliar de mercerizado

Producto que sirve para mejorar el poder humectante de las lejías de mer-cerizado y para acelerar así su pene-tración uniforme en las fibras.

Observación: Se trata de agentes humectantes estables en lejías fuertemente concentradas, que contienen un componente con eficacia tensioactiva y emulsionante en la lejía (alquilsulfatos de bajo peso molecular, aceites altamente sulfatados, cresoles, xilenoles) y una substancia antiespumante y humectante, insoluble por sí misma en la lejía, pero solubilizada por hidrotropía (por ejemplo: butilglicol, aminas etoxiladas, etc.).

it —

ne merceriseerloogbevochtiger

Produkt welke tot doel heeft de effekti-viteit en snelheid van het merceriseren te verbeteren.

pl pomocniczy środek merceryzacyjny

Produkt służący do poprawy zdolności zwilżania ługów merceryzacyjnych, przyśpieszający dzięki temu ich równo-mierną penetrację w głąb włókien.

Uwaga: Produktami takimi są środki zwilżające trwałe w ługach o wysokim stężeniu. Oparte są one na środkach, które działają w ługach zarówno jako związki powierzchniowo czynne oraz jako emulgatory (niskocząsteczkowe alkilosiarczany, wysokosulfonowane oleje, kreozole, ksylenole) oraz na substancji przeciwpieniącej i zwilżającej (np. glikol butylowy, oksyetylenowane aminy itd.) rozpuszczonej na zasadzie hidrotropii.

392 metallic dye
fr colorant métallique
de Metallhaarfärbemittel
el colorante metálico
it colorante metallico
ne metaalhoudende kleurstof
pl barwnik metalizowany, barwnik metalo-kompleksowy

393 micelle

In the special case of solutions of surface active agents: an aggregate made up of molecules and/or ions, forming above a certain critical concentration.

fr micelle

Dans le cas particulier des solutions d'agents de surface: agrégat organisé de molécules et/ou d'ions se formant au-dessus d'une certaine concentration critique.

de Micelle

In dem speziellen Fall der Lösungen grenzflächenaktiver Verbindungen: ge-ordnete Aggregation von Molekülen und/oder Ionen, die sich oberhalb einer gewissen kritischen Konzentration bildet.

el micela

En el caso particular de una disolución de agente de superficie: agrupación organizada de moléculas o iones, que se forma a concentraciones superiores a una concentración crítica.

it micella

Nel caso particolare delle soluzioni di tensioattivi: aggregato organizzato di molecole e/o ioni che si forma a concen-trazione superiore ad un certo valore detto critico.

ne micel

In het bijzondere geval van oplossingen van oppervlakaktieve stoffen: aggregaat bestaande uit moleculen en/of ionen dat zich vormt boven een bepaalde kri-tische concentratie.

pl micela

W szczególnym przypadku związków powierzchniowo czynnych jest to upo-rządkowana agregacja cząsteczek lub jonów, powstająca powyżej określonego stężenia krytycznego.

394 micelle formation
fr formation de micelles
de Micellbildung
el formación de micelas
it formazione di micelle
ne micel-vorming
pl powstawanie miceli

395 middle soap, clotted soap†

The phase of soap in the pan, anisotropic, translucent and very viscous, nematic in structure, of lower concentration than finished soap, appearing as thick clots or compact relatively immobile masses, rubbery in consistency, taking up almost the entire volume of the pan. This phase, which is highly undesirable in manufac-ture, appears if the electrolyte content falls below a certain value. Middle soap is difficult to redissolve and its appear-ance during manufacture is a regrettable accident.

fr savon médian, savon en grumeaux
Phase du savon en chaudière, anisotrope,
translucide et très visqueuse, à structure
nématique, de concentration inférieure
à celle du savon lisse, se présentant en
grumeaux épais ou en masses compactes
de très faible mobilité, à consistance
élastique, pouvant occuper presque tout
le volume de la chaudière. Cette phase,
hautement indésirable en fabrication,
apparaît si la teneur en électrolytes tombe
au-dessous d'une certaine valeur. Le
savon médian est difficile à redissoudre
et son apparition constitue un accident
redoutable en cours de fabrication.

de Klumpseife, Mittelseife
Anisotrope, transparente, hochviskose
Phase nematischer Struktur, von niedrigerer Konzentration als der geschliffene
Seifenkern. Diese Phase tritt zunächst
in Form zähflüssigen elastischen Klum-
pen auf, kann sich aber mitunten auf
den gesamten Inhalt des Kessels erstrek-
ken. Diese Phase bildet sich, wenn der
Elektrolytegehalt unter einen bestimmten
Wert fällt. Die Klumpseife ist schwer
wieder in Lösung zu bringen und bewirkt
eine schwerwiegende Störung des Her-
stellungsprozesses.

el jabón medio, jabón en grumos
Fase transparente y muy viscosa, de
estructura nemática, que se presenta
primero en forma de grumos, pero que
durante la cocción puede llegar a ocupar
todo el contenido de la caldera. La formación de jabón medio constituye una
grave alteración de la fabricación.
Observación: Estado nemático. Estado de las
fases cristalino-líquidas en el cual las moléculas
o partículas anisótropas están regularmente
orientadas en una dirección. En las direcciones
restantes, las moléculas se orientan al azar.

it sapone mediano, sapone in grumi
La fase del sapone in caldaia, a carattere
anisotropo translucido e molto viscoso,
di concentrazione più bassa rispetto al
sapone finito, di aspetto grumoso e in
masse compatte e che occupa quasi
l'intero volume della caldaia. Questa
fase, che provoca seri inconvenienti
nelle lavorazioni, compare se il contenuto di elettrolita è al di sotto di certi
valori; il sapone in grumi è di difficile
ridissoluzione e produce notevoli inconvenienti durante le lavorazioni.

ne bonkzeep, middenzeep
Anisotrope, transparente, zeer visceuse
fase met nematische struktuur, van
lagere concentratie dan de afgemaakte
kernzeep. Deze fase treedt eerst in de
vorm van taaie elastische klompen op,
die tenslotte praktisch de gehele ketel-
inhoud kan beslaan. Deze fase, welke in
hoge mate ongewenst is, vormt zich
als het electrolytgehalte onder een
bepaalde waarde daalt. Bonkzeep is
moeilijk weer in oplossing te brengen
en veroorzaakt een ernstige storing
tijdens de fabrikatie.

pl mydło zgęstniałe
Anizotropowa, przeźroczysta i wysoko-
lepka faza mydła w kotle warzelnym,
o nematycznej budowie i niższym stężeniu niż wysół płynny, pojawiająca się
w formie grubych grudek lub zwartych,
stosunkowo nieruchliwych, gumowatych
w konsystencji bryłek, zajmująca prawie
całą objętość kotła. Ta wysoce niepożą-
dana przy produkcji mydła faza pojawia
się wtedy, gdy zawartość elektrolitu
spadnie poniżej pewnej wartości. Mydło
zgęstniałe jest trudne do ponownego
rozpuszczenia i powoduje kłopotliwe
zakłócenia procesu produkcyjnego.
† See appendix

396 middle soap formation
The undesirable formation of soap in
clots, as a result of too low concentration
of electrolytes, either from saponification
with insufficient alkali or excessive dilution with water. This formation produces
undue thickening of the soapy mass
which becomes difficult to handle or
redissolve.

fr formation de savon médian
Formation indésirée de savon en grumeaux, comme conséquence d'une diminution de la concentration d'électrolytes,
soit par saponification avec une quantité
insuffisante d'alcali, soit par dilution
trop poussée avec de l'eau. Cette formation produit un fort épaississement
de la masse savonneuse, qui devient
difficile à manier et à redissoudre.

de Zusammenfahren
Unerwünschte Bildung von Klumpseife
infolge Verringerung der Elektrolyt-
Konzentration in der Seifenmasse,
bedingt durch Verseifung mit ungenügen-

der Alkalimenge oder durch zu starke Verdünnung mit Wasser. Durch das Zusammenfahren nimmt die Masse eine sehr steife Konsistenz, wird schwer behandelbar und ist schwer wieder in Lösung zu bringen.

el formación de jabón medio
Formación indeseada de jabón en grumos, como consecuencia de una disminución de la concentración de electrólitos en la masa de jabón sea por falta de álcali, o por una excesiva dilución con agua. A causa de esta formación, la masa toma la consistencia de una pasta espesa, que no puede ser bombeada y que es difícil de redisolver.

it formazione di sapone mediano
L'indesiderata formazione di sapone in grumi, quale risultato sia dell'uso di elettroliti a concentrazione troppo bassa, sia dalla saponificazione in difetto di alcali caustico, o eccessiva diluizione con acqua. Questa formazione produce un indesiderato ispessimento della massa saponosa, che diventa difficile da maneggiare e da ridisciolgiere.

ne bonk-vorming, vorming van middenzeep
Ongewenste vorming van bonken door verlaging van de electrolytconcentratie in de zeepmassa veroorzaakt door verzeping met onvoldoende hoeveelheid alkali of door te sterke verdunning met water. Hierdoor krijgt de massa een zeer stijve konsistentie, wordt moeilijk te verwerken en in oplossing te brengen.

pl tworzenie się mydła zgęstniałego
Niepożądane tworzenie się mydła zgęstniałego w postaci grudek, będące wynikiem zbyt niskiego stężenia elektrolitów, spowodowanego bądź przez prowadzenie zmydlania przy pomocy niewystarczającej ilości alkaliów, bądź przez nadmierne rozcieńczenie wodą. Proces ten powoduje nadmierne zgęstnienie masy mydlanej, która staje się trudna do obróbki lub ponownego rozpuszczenia.

397 moderately hard water
fr eau moyennement dure
de mässig hartes Wasser
el agua moderadamente dura, agua medianamente dura
it acqua di durezza media
ne matig hard water
pl woda umiarkowanie twarda

398 monomolecular layer, monolayer
Adsorption layer which, under determined conditions of concentration, is limited to unit molecular thickness of a surface active agent.
fr couche monomoléculaire
Couche d'adsorption qui, dans des conditions déterminées de concentration, se limite à l'épaisseur unitaire de molécules d'un agent de surface.
de monomolekulare Schicht
Adsorptionsschicht, welche unter bestimmten Konzentrationsbedingungen auf die Dicke eines Tensidmoleküls beschränkt ist.
el capa monomolecular, monocapa
Capa de adsorción que, en determinadas condiciones de concentración, se limita al espesor de una molécula del agente de superficie.
it strato monomolecolare
Strato d'assorbimento il cui spessore, per particolari condizioni di concentrazione, è limitato alle dimensioni di una molecola di tensioattivo.
ne mono-moleculaire laag
Adsorptielaag die onder bepaalde kondities t.a.v. de concentratie een dikte heeft van één molecuul van de oppervlakaktieve stof.
pl warstwa monomolekularna, warstwa jednocząsteczkowa
Warstwa adsorpcyjna, która w określonych warunkach stężenia ograniczona jest do grubości jednej cząsteczki związku powierzchniowo czynnego.

399 mottled soap
Soap obtained from the lye–nigre mixture during cooling when the viscosity of the mass prevents the sedimentation of the nigre. This remains in the finished soap in the form of coloured streaks if pigment has previously been added to the soap.
fr savon marbré
Savon obtenu d'une démixtion "lissegras", ayant lieu dans les mises, pendant le refroidissement, lorsque la viscosité de la masse ne permet pas la sédimentation du gras. Celui-ci reste mélangé à la phase savon lisse sous forme de nervures colorées, si au préalable on a ajouté un pigment au savon.
de marmorierte Seife
Seife, die beim Abkühlen durch die in

der Seifenform stattfindende Entmischung—"geschliffener Kern-Leimniederschlag" —entsteht. Diese teilweise Entmischung verursacht die Bildung transparenter oder—bei Farbstoffzusatz—farbiger Adern in der Seife.

el jabón veteado, jabón de pintas
Jabón obtenido por una separación "liso–bajos" que tiene lugar en los moldes durante el enfriamiento. La separación parcial produce la formación de vetas transparentes o coloreadas, estas últimas si se han añadido pigmentos.

it sapone marmorato
Sapone ottenuto dalla miscela di collette e liscivie durante il raffreddamento, quando la viscosità della massa impedisce la separazione delle collette. Queste rimangono nel sapone finito in forma di strati colorati, se un pigmento è stato precedentemente aggiunto al sapone.

ne gemarmerde zeep
Zeep, die bij het afkoelen uit het looglijmzeep mengsel ontstaat, als viskositeit van de massa het afscheiden van de lijmzeep verhindert. Deze blijft in de afgemaakte zeep in de vorm van gekleur-

de aders, als aan de zeep vooraf pigment toegevoegd is.

pl mydło marmurkowe
Mydło otrzymane z mieszaniny wysołu płynnego z klejem mydlanym podczas schładzania, kiedy to lepkość masy mydlanej zabezpiecza układ przed sedymentacją kleju mydlanego. Klej mydlany pozostaje w wysole płynnym w postaci barwnych pasemek, jeśli uprzednio do mydła był dodany barwnik.

400 mouth freshener
fr rafraîchissant pour l'haleine
de Mittel zur Erfrischung des Atems
el refrescante bucal
it rinfrescante per la bocca
ne middel tot verfrissen van de adem
pl dezodorant do ust

401 mouth wash
fr bain de bouche, eau dentifrice, collutoire
de Mundwasser, Mundspülwasser
el lavado de boca, colutorio
it lavaggio della bocca, colluttorio
ne mondwater, mondspoelmiddel
pl woda do ust, woda do płukania ust

N

402 nail bleach
fr produit pour blanchir les ongles
de Nagelbleichmittel
el producto para blanquear las uñas
it prodotto per schiarire le unghie
ne middel om nagels te bleken
pl środek do wybielania paznokci

403 nail cream
fr crème pour les ongles
de Nagelcreme
el crema para las uñas
it crema per le unghie
ne nagelcrème
pl krem do paznokci

404 nail drier
fr séchoir à ongles
de Nageltrockner
el secador para las uñas
it essiccativi per unghie
ne nageldroger
pl środek osuszający do paznokci

405 nail enamel
fr vernis à ongles
de Nagelglasur
el esmalte de uñas
it vernice per le unghie
ne nagelvernis
pl emalia do paznokci

406 nail lacquer
fr laque à ongles
de Nagellack
el laca de uñas
it smalto per le unghie
ne nagellak
pl lakier do paznokci

407 nail paste
fr pâte pour les ongles
de Nagelpaste
el pasta para las uñas
it pasta per le unghie
ne nagelpasta
pl pasta do paznokci

408 nail polish powder
fr poudre à polir les ongles
de pulverförmige Nagelpolitur
el polvos para pulir las uñas
it polvere per pulire le unghie
ne poedervormige nagelpolitoer
pl proszek do polerowania paznokci

409 (nail) top coat
fr protecteur de vernis
de abdeckender Überzug
el capa protectora
it protettore di vernice
ne toplaag
pl warstwa ochraniająca (emalię do paznokci)

410 nail whitener
fr produit pour éclaircir les ongles
de Nagelweiss-Präparat
el producto abrillantador de las uñas
it prodotto per far risplendere le unghie
ne middel om nagels te witten
pl preparat do rozjaśniania paznokci

411 natural colour of the hair
fr teinte naturelle du cheveu
de natürliche Haarfarbe
el tono natural del cabello
it tinta naturale del capello
ne natuurlijke kleur van het haar
pl naturalna barwa włosów

412 natural shade
fr couleur naturelle, teinte naturelle
de Naturfarbe
el color natural, tono natural
it tinta naturale
ne natuurlijke schakering
pl naturalny odcień

413 natural soil
fr salissure naturelle
de natürlicher Schmutz
el suciedad natural
it sporco naturale
ne natuurlijk vuil
pl brud naturalny

414 neutralizer
fr neutralisant
de Fixiermittel
el neutralizante
it neutralizzante
ne neutraliseermiddel
pl środek zobojętniający

415 neutralizing solution
fr solution neutralisante
de neutralisierende Lösung
el solución neutralizante
it soluzione neutralizzante
ne neutraliserende oplossing
pl roztwór zobojętniający

416 new growth of hair
fr repousse
de Nachwuchs (des Haares)
el recrecimiento del cabello
it recrescita dei capelli
ne opnieuw groeien van het haar
pl odrastanie włosów

417 night cream
fr crème de nuit
de Nachtcreme
el crema de noche
it crema per notte
ne nachtcrème
pl krem na noc

418 nigre
An isotropic solution of soap containing electrolytes, which is separated from the finished soap by decantation or centrifuging after the finishing process.
fr gras
Solution isotrope de savon, contenant des électrolytes, qui se sépare du savon lisse par décantation ou centrifugation après l'opération de liquidation.
de Leimniederschlag
Isotrope, elektrolythaltige Seifenlösung, die sich von dem geschliffenen Seifenkern durch Absetzenlassen oder Zentrifugieren nach dem Ausschleifen abtrennen lässt.
el bajos
Disolución de jabón que contiene electrólitos, que se separa del jabón liso por decantación o centrifugación después de la operación de liquidación.
it collette
Una soluzione isotropa di sapone contenente elettroliti, che si separa dal sapone finito a mezzo decantazione o centrifugazione, dopo il processo di rifinitura.
ne lijmzeep (bodem)
Een isotrope, electrolyt bevattende zeepoplossing die van de afgemaakte kernzeep verwijderd wordt door decantatie of centrifugeren na het afmaken.
pl klej mydlany, klej pomydlany, klej spodni
Izotropowy roztwór mydła zawierający elektrolity, wydzielony z wysołu płynnego przez dekantację lub wirowanie po zakończeniu procesu wykańczania.

419 nitro dyestuff
fr colorant nitré
de Nitrofarbstoff
el colorante nitrado
it colorante nitrato
ne nitrokleurstof
pl barwnik nitrowy

420 non-biodegradable surface active agent
Surface active agent which resists biodegradation.
fr agent de surface biorésistant
Agent de surface qui résiste à la biodégradation.
de Biologisch "hartes" Tensid
Tensid, welches einem biologischen Abbau nur in unzureichendem Masse unterliegt.
el agente de superficie bio-resistente
Agente de superficie que resiste a la biodegradación.
it tensioattivo bioresistente
Tensioattivo che resiste alla biodegradazione in presenza di acqua.
ne niet biologisch afbreekbare oppervlakaktieve stof
Oppervlakaktieve stof die in te geringe mate een verandering ondergaat door biodegradatie.
pl nierozkładalny biologicznie związek powierzchniowo czynny
Związek powierzchniowo czynny odporny na rozkład biologiczny.

421 non dusting, non sneezing
fr non poussiéreux
de nichtstaubend, staubfrei
el no pulverulento
it non polveroso
ne stofvrij
pl niepylący, bezpylny

422 non-greasy cream
fr crème non grasse
de fettfreie Creme
el crema no grasa
it crema non grassa
ne niet vette crème
pl krem nietłusty

423 non-ionic surface active agent
A surface active agent which does not produce ions in an aqueous solution. The solubility in water of non-ionic surface active agents is due to the presence in the molecules of functional groups which have a strong affinity for water.

fr agent de surface non ionique
Agent de surface ne donnant pas naissance à des ions en solution aqueuse. La solubilité dans l'eau des agents de surface non ioniques est due à la présence dans leurs molécules de groupements fonctionnels ayant une forte affinité pour l'eau.

de nichtionische grenzflächenaktive Verbindung
Grenzflächenaktive Verbindung, die in wässriger Lösung keine Ionen bildet. Die Wasserlöslichkeit der nichtionogenen grenzflächenaktiven Verbindungen wird dadurch bedingt, dass in ihren Molekülen funktionelle Gruppen vorhanden sind, die eine starke Affinität zu Wasser haben.

el agente de superficie no iónico
Agente de superficie que no produce iones en disolución acuosa. La solubilidad en agua de los agentes de superficie no iónicos se debe a la presencia en sus móléculas de grupos funcionales que poseen una fuerte afinidad para el agua.

it tensioattivo non-ionico
Tensioattivo che non dà ioni in soluzione acquosa. La solubilità nell'acqua dei tensioattivi non-ionici è dovuta alla presenza nelle loro molecole di gruppi funzionali aventi una forte affinità per l'acqua.

ne niet-ionogene aktieve stof
Een oppervlakaktieve stof die in waterige oplossingen geen ionen produceert. De oplosbaarheid in water van niet-ionogene aktieve stoffen wordt veroorzaakt door het feit dat de moleculen funktionele groepen bevatten die een sterke affiniteit voor water hebben.

pl niejonowy związek powierzchniowo czynny
Związek powierzchniowo czynny nie tworzący jonów w roztworze wodnym. Rozpuszczalność niejonowego związku powierzchniowo czynnego w wodzie jest wynikiem obecności w cząsteczkach grup funkcyjnych, charakteryzujących się silnym powinowactwem do wody.

424 nonionics
fr composés non-ioniques
de nichtionogene Verbindungen

el compuestos no iónicos
it composti non ionici
ne nonionaktieve stoffen
pl niejonowe związki (powierzchniowo czynne)

425 non-polar group
The organic part of the molecule, in which the distribution of electrons does not cause a considerable electrical dipole moment. Such a group conditions the affinity for organic solvents of low polarity and consequently the lipophilic character of the molecule.

fr radical apolaire
Partie organique de la molécule dont les caractéristiques de répartition électronique n'entraînent pas un moment électrique dipolaire notable. Un tel radical conditionne l'affinité pour les solvants organiques de faible polarité et par suite le caractère lipophile de la molécule.

de apolarer Rest
Organischer Teil des Moleküls, dessen Elektronen-Verteilung keinen nennenswerten Beitrag zum Dipolmoment liefert. Ein solcher Rest bedingt die Affinität zu organischen Lösungsmitteln geringer Polarität und infolgedessen den lipophilen Charakter des Moleküls.

el radical apolar
Parte orgánica de la molécula en la que la distribución de electrones no produce un momento eléctrico dipolar apreciable. Este radical ocasiona la afinidad para los disolventes orgánicos de polaridad débil y da, por tanto, el carácter lipófilo a la molécula.

it radicale apolare
Parte organica della molecola avente distribuzione elettronica tale da non contribuire sensibilmente al momento elettrico dipolare. Questo radicale determina l'affinità per solventi organici di debole polarità e quindi il carattere lipofilo della molecola.

ne niet-polaire groep
Het organische deel van het molecuul waarvan de spreiding van de elektronen geen belangrijk elektrisch dipoolmoment met zich meebrengt. Een dergelijke groep bepaalt de affiniteit voor organische stoffen met een lage polariteit, dus het lipofiele karakter van het molecuul.

pl grupa niepolarna
Organiczna część cząsteczki, w której rozkład elektronów nie powoduje wytworzenia znaczniejszego momentu dipolowego. Grupa taka warunkuje powinowactwo do rozpuszczalników organicznych o niskiej polarności, a więc w konsekwencji lipofilowy charakter cząsteczki.

426 nourishing cream
fr crème nutritive
de Nährcreme
el crema nutritiva
it crema nutriente
ne voedingscrème
pl krem odżywczy

O

427 oil emulsion (symbol H-L: water in oil[1])
An emulsion in which the continuous
phase is a liquid insoluble in water.

fr émulsion de type huileux (symbole
H-L: eau dans l'huile[2])
Emulsion dont la phase continue est un
liquide insoluble dans l'eau.

de ölige Emulsion (Abkürzung H-L =
Wasser in Öl[3])
Emulsion, deren kontinuierliche Phase
eine in Wasser unlösliche Flüssigkeit ist.

el emulsión oleosa
Emulsión, cuya fase continua es un
líquido insoluble en agua.

it emulsione di tipo oleoso
Emulsione la cui fase continua è costituita
da un liquido insolubile in acqua.

ne emulsie in olie (symbool H-L: water
in olie[4])
Emulsie waarin een in water onoplosbare
vloeistof de kontinue fase vormt.

pl emulsja olejowa (symbol H-L: woda w
oleju[5])
Emulsja, w której fazą rozpraszającą jest
ciecz nierozpuszczalna w wodzie.

[1] From the Greek: L = lipos, H = hydor.
[2] Du grec: L = lipos, H = hydor.
[3] Vom griechischen L = lipos, H = hydor.
[4] Van het grieks: L = lipos, H = hydor.
[5] Z greckiego L = lipos, H = hydor.

428 omega phase
The crystalline form, stable above 70°C,
always appearing in solid finished soap
above this temperature. At a lower
temperature, and at ambient temperature, the omega phase in a metastable
form can be obtained if the finished soap
is cooled rapidly and without agitation.
By strong mechanical action on this
metastable state, the omega phase is
converted into the beta phase which is
stable in the cold. Soaps in the omega
phase have poor foaming properties
because this phase has the lowest solution
rate. They are, moreover, less firm than
soaps in the beta phase.

fr phase oméga
Forme cristalline, stable au-dessus de
70°C, se retrouvant toujours dans le
savon lisse solidifié au-dessus de cette
température. A température plus basse
et même à la température ordinaire, on
peut obtenir la phase oméga à l'état
métastable si on refroidit le savon lisse
rapidement et sans agitation. Par action
mécanique énergique sur cet état métastable, la phase oméga se transforme
en phase bêta, stable à froid. Les savons
en phase oméga moussent mal, parce
que cette phase a la vitesse de dissolution
la plus basse. Ils sont aussi moins fermes
que les savons en phase bêta.

de Omega-Phase
Kristalline, über 70°C stabile Form,
die immer im festen, geschliffenen Seifenkern oberhalb dieser Temperatur enthal-
ten ist. Sie kann auch in metastabilem
Zustand bei niedrigerer und sogar
Raumtemperatur vorkommen, voraus-
gesetzt, dass man den geschliffenen
Seifenkern schnell und ohne mechanische
Einwirkung abkühlt. Durch starke me-
chanische Einwirkung auf die Seife in
dieser Form geht die Omega-Phase in die
Beta-Phase über, die in der Kälte die
stabile Form ist. Die Seifen der Omega-
Phase schäumen schlecht an, da diese
Phase die niedrigste Lösungsgeschwindig-
keit hat. Sie sind weicher als Seifen in
Form der Beta-Phase.

el fase omega
Forma cristalina, estable por encima de
70°C, que se encuentra siempre en el
jabón liquidado, solidificado por encima
de esta temperatura. A temperatura más
baja y aun a la temperatura ordinaria,
puede obtenerse la fase omega en estado
metaestable, si se enfría el jabón liquidado rápidamente y sin agitación. Por
acción mecánica sobre este estado metaestable, la fase omega se transforma en
fase beta, estable en frío. Los jabones
en fase omega producen poca espuma,
porque esta fase posee la más baja
velocidad de disolución. Son también
más blandos que los jabones en fase
beta.

it fase omega
Forma cristallina, stabile oltre i 70°C,
che appare sempre in saponi finiti solidi
al disopra di questa temperatura. A
temperatura più bassa e a temperatura
ambiente, la fase omega, in forma
metastabile, può essere ottenuta se il
sapone finito è raffreddato rapidamente

senza agitazione. Con energica azione meccanica sulla fase metastabile, la fase omega passa alla fase beta, che è stabile a freddo. I saponi della fase omega hanno basse proprietà schiumogene perchè questa fase ha la solubilità più bassa. Inoltre essi sono meno compatti dei saponi della fase beta.

ne omega-fase
Kristallijnen, boven 70°C stabiele vorm, die steeds in de vaste, afgemaakte kernzeep boven deze temperatuur voorhanden is. Zij kan ook in metastabiele toestand bij lagere en zelfs bij kamertemperatuur voorkomen, indien men de afgemaakte kernzeep snel en zonder mechanische bewerking afkoelt. Door sterke mechanische bewerking gaat de zeep in deze vorm van de omega-fase in beta-fase over, welke laatste de bij lage temperatuur stabiele vorm is. De zepen van de omega-fase schuimen slecht aan, daar deze fase de laagste oplosnelheid bezit. Zij zijn weker dan zepen in de beta-fase.

pl faza omega
Forma krystaliczna, trwała w temperaturach powyżej 70°C, zawsze występująca w stałym, wykończonym mydle powyżej tej temperatury. W temperaturach niższych oraz w temperaturze pokojowej faza omega może być otrzymana w formie metatrwałej, jeśli wykończone mydło zostanie gwałtownie schłodzone bez obróbki mechanicznej. Faza omega przeprowadzana jest w trwałą na zimno fazę beta poprzez energiczną mechaniczną obróbkę tej metatrwałej formy. Mydła w fazie omega odznaczają się słabymi własnościami pieniącymi, ponieważ faza ta posiada najmniejszą szybkość rozpuszczania. Ponadto mydła te są miększe niż mydła w fazie beta.

429 optical whitening agent, optical bleaching agent, fluorescent whitening agent, optical brightener, optical dye, brightening agent
fr azurant optique, agent de blanchiment optique, agent blanchissant optique, azureur

de optischer Aufheller, optisches Bleichmittel, Weisstöner, Fluoreszenzfarbstoff
el agente blanqueante óptico, blanqueador óptico
it azzurrante ottico, sbiancante ottico, agente di sbianca ottica, agente fluorescente
ne optisch bleekmiddel, optisch witmiddel
pl rozjaśniacz optyczny

430 oral tissue
fr tissus buccaux
de Mundschleimhaut
el tejidos bucales
it tessuti orali
ne mondweefsel
pl błona śluzowa ust

431 oxidation
fr oxydation
de Oxydation
el oxidación
it ossidazione
ne oxydatie
pl utlenianie

432 oxidation hair dye
fr colorant d'oxydation pour les cheveux, teinture d'oxydation
de Oxydationsfarbstoff
el colorante de oxidación
it colorante di ossidazione
ne oxydatieve haarverf
pl utleniający barwnik do włosów

433 oxidizing rinse
fr rinçage oxydant
de oxydierende Spülung
el enjuague oxidante, lavado oxidante
it risciacquo ossidante
ne oxyderend spoelen
pl płukanka utleniająca

434 oxygenated bleaching compound
fr agent de blanchiment oxygéné
de Sauerstoffbleichmittel
el agente de blanqueo oxigenado, agente de blanqueo con oxígeno
it agente di sbianca ossigenato
ne zuurstofbevattend bleekmiddel
pl środek bielący tlenowy, utleniający związek bielący

P

435 pan room
fr savonnerie
de Siederei
el jabonería
it saponificio, reparto caldaie
ne ziederij, zeepziederij
pl warzelnia

436 papilla
fr papille
de Papille
el papila
it papilla
ne papil
pl brodawka

437 particle size
fr dimension granulométrique, grosseur des billes
de Teilchengrösse, Korngrösse
el dimensiones granulométricas, espesor de los granos
it diametro delle sfere
ne deeltjesgrootte
pl wielkość ziarna, uziarnienie

438 particle size analysis, determination of particle size distribution, measurement of the particle size characteristics
fr analyse granulométrique
de Bestimmung der Korngrösse
el análisis granulométrico
it analisi granulometrica
ne bepaling van de deeltjesgrootte
pl analiza rozkładu wielkości ziarn, analiza granulometryczna

439 particle size distribution
fr distribution granulométrique
de Korngrössenbereich, Korngrössenverteilung
el distribución granulométrica
it distribuzione granulometrica
ne verdeling van de deeltjesgrootte
pl rozkład wielkości ziarn

440 pasting
The conversion of fatty matter into an isotropic solution of soap by saponification.
fr empâtage
Passage des matières grasses par saponification à l'état de solution isotrope de savon.

de Verleimung
Durch Verseifung bewirkter Übergang der Fettstoffe in den Zustand isotroper Seifenlösung (Seifenleim).
el empaste
Disminución de la concentración de electrólitos después de la saponificación, hasta por debajo de la concentración de la lejía límite, lo que tiene como consecuencia el paso del jabón al estado de disolución isótropa.
it impasto
Il passaggio degli acidi grassi ad una soluzione isotropa di sapone a mezzo saponificazione.
ne verlijming
Omzetting van vetten tot een isotrope zeepoplossing door verzeping.
pl wytwarzanie kleju mydlanego
Przejście substancji tłuszczowej w wyniku zmydlenia w izotropowy roztwór mydła.

441 patch test
fr patch-test
de Läppchentest, Läppchenprobe
el patch-test
it prova del panno
ne lapjesproef
pl próba płatkowa

442 penetrating power
fr pouvoir pénétrant
de Eindringungsvermögen
el poder penetrante
it potere penetrante
ne indringend vermogen
pl zdolność penetracji, zdolność wnikania

443 peptide linkage
fr liaison peptidique
de Peptidbindung
el enlace péptido
it legame peptidico
ne peptide-band
pl wiązanie peptydowe

444 peptization
The formation of a stable dispersion from flocs or aggregates.
fr peptisation
Formation d'une dispersion stable à partir de flocons ou agrégats.
de Peptisation
Bildung einer stabilen Dispersion aus Flocken oder aus Aggregaten.

el peptización
    Formación de una dispersión estable,
    a partir de flóculos o de agregados.
it peptizzazione
    Formazione di una dispersione stabile
    ottenuta da aggregati grossolani.
ne peptiseren
    De vorming van een stabiele dispersie uit
    vlokken of aggregaten.
pl peptyzacja
    Tworzenie trwałej dyspersji z kłaczków
    lub agregatów

445 **peptizing agent, peptizer**
    A substance capable of promoting
    peptization.
fr agent peptisant, peptisant
    Produit apte à promouvoir la peptisa-
    tion.
de Peptisiermittel
    Produkt, das in der Lage ist, die Peptisa-
    tion zu fördern.
el agente peptizante, peptizante
    Producto capaz de promover la pepti-
    zación.
it peptizzante
    Sostanza atta a promuovere la peptizza-
    zione.
ne peptiseermiddel
    Een stof die peptiseren mogelijk maakt.
pl środek peptyzujący
    Substancja zdolna do pobudzenia procesu
    peptyzacji.

446 **permanent wave**
fr ondulation permanente
de Dauerwelle, Dauerverformung
el ondulación permanente
it ondulazione permanente
ne haargolf
pl trwała ondulacja

447 **permanent waving**
fr permanente, indéfrisable
de Dauerwellen
el ondulación permanente, permanente
it permanente
ne onduleren
pl ondulowanie na trwało

448 **permanent waving of the heated type**
fr permanente chaude, permanente à chaud
de Heissdauerwelle
el permanente en caliente
it permanente a caldo
ne onduleren bij verhoogde temperatuur
pl trwała ondulacja na gorąco

449 **persalt**
fr persel
de Persalz
el persales
it persale
ne perzout
pl sól kwasu nadtlenowego, nadsól

450 **personal washing agent**
fr détersif pour la peau, détergent pour
    les soins corporels
de Hautreinigungsmittel
el detergente para la piel, detergente para
    uso personal
it detergente per la pelle, detergente per
    uso personale
ne huidreinigingsmiddel
pl środek do mycia ciała

451 **phosphation**
    In the particular case of surface active
    agents, chemical reaction giving rise to
    the formation of phosphoric esters.
fr phosphatation
    Dans le cas particulier des agents de
    surface, réaction chimique permettant
    d'obtenir des esters phosphoriques.
de Phosphatierung
    Im speziellen Fall der grenzflächenakti-
    ven Körper: chemische Reaktion, bei
    der Phosphorsäureester entstehen.
el fosfatación
    En el caso particular de los agentes de
    superficie: reacción química que permite
    la obtención de ésteres fosfóricos.
it fosfatazione
    Nel caso particolare dei tensioattivi:
    reazione chimica che porta alla forma-
    zione di esteri fosforici.
ne fosfatering
    In het speciale geval van oppervlakaktieve
    stoffen is dit de chemische reactie waarbij
    fosforzure esters worden gevormd.
pl fosforowanie
    W szczególnym przypadku związków
    powierzchniowo czynnych reakcja pro-
    wadząca do utworzenia estrów kwasu
    fosforowego.

452 **phosphonation**
    Chemical reaction or sequence of chemi-
    cal reactions, leading to the introduction
    into a molecule of one or more phos-
    phonic radicals by direct carbon/phos-
    phorus linkage.

fr phosphonation
Réaction chimique, ou suite de réactions chimiques, permettant d'introduire dans une molécule une ou plusieurs fonctions phosphoniques par liaison directe carbone/phosphore.

de Phosphonierung
Chemische Reaktion oder Reaktionsfolge, bei der in ein Molekül eine oder mehrere Phosphorsäuregruppen eingeführt werden mit direkter Bindung Phosphor/Kohlenstoff.

el fosfonación
Reacción química, o serie de reacciones químicas, que permite la introducción de una o de varias funciones fosfónicas por enlace directo carbono–fósforo.

it fosfonazione
Reazione chimica (o serie di reazioni chimiche) che introduce nella molecola una o più radicali fosfonici, con legame diretto carbonio–fosforo.

ne fosfonering
Chemische reactie of reeks van chemische reacties waarbij een of meer fosforzuurradicalen via een direkte koolstof–fosforverbinding in een molecuul worden ingevoerd.

pl fosfonowanie
Reakcja chemiczna lub ciąg reakcji chemicznych, w których do cząsteczki wprowadzona zostaje jedna lub więcej reszt kwasu fosforowego poprzez bezpośrednie wiązanie węgla z fosforem.

453 pigmentation
fr pigmentation
de Pigmentierung
el pigmentación
it pigmentazione
ne pigmentatie
pl pigmentacja

454 plastic container
fr emballage plastique
de Plastikbehälter
el recipiente de plástico, envase de plástico
it imballaggio in plastica
ne houder uit kunststof
pl pojemnik z tworzywa sztucznego

455 plodder
fr peloteuse-boudineuse
de Strangpresse
el compresora, embutidora
it trafila

ne stangenpers
pl peloteza

456 polar group
A functional group, in which the distribution of electrons tends to give a considerable electrical dipole moment to the molecule. Such a group conditions the affinity for markedly polar surfaces, the affinity for water in particular and the hydrophilic character of the molecule.

fr groupement polaire
Groupement fonctionnel dont les caractéristiques de répartition électronique tendent à assurer à la molécule un moment électrique dipolaire notable. Un tel groupement conditionne l'affinité pour les surfaces nettement polaires, pour l'eau en particulier, et le caractère hydrophile de la molécule.

de polare Gruppe
Funktionelle Gruppe, deren Elektronenverteilung dahin zielt, dem Molekül ein beträchtliches elektrisches Dipolmoment zu verleihen. Eine solche Gruppe bedingt die Affinität zu ausgesprochen polaren Oberflächen, im besonderen die Affinität zu Wasser, und den hydrophilen Charakter des Moleküls.

el grupo polar
Grupo funcional en el que la distribución de electrones tiende a producir en la molécula un momento eléctrico dipolar apreciable. Este grupo ocasiona la afinidad para las superficies francamente polares, especialmente para el agua, y da a la molécula su carácter hidrófilo.

it gruppo polare
Gruppo funzionale avente distribuzione elettronica tale da fornire alla molecola un marcato momento elettrico di-polare. Tale gruppo determina l'affinità per le superfici marcatamente polari, in particolare per l'acqua, e il carattere idrofilo della molecola.

ne polaire groep
Een funktionele groep die door de elektronenverdeling de neiging vertoont, het molecuul een vrij groot dipoolmoment te geven. Een dergelijke groep bepaalt de affiniteit voor uitgesproken polaire vlakken, in het bijzonder de affiniteit voor water, en het hydrofiele karakter van het molecuul.

pl grupa polarna

Grupa funkcyjna cząsteczki, w której rozkład elektronów przyczynia się do nadania jej znacznego momentu dipolowego. Grupa taka warunkuje powinowactwo do wyraźnie polarnych powierzchni w szczególności do wody oraz hydrofilowy charakter cząsteczki.

457 polar-non-polar structure
The structure of a molecule which has at least one polar group and a large nonpolar group. Such a structure conditions the hydrophilic and lipophilic characters of the molecule.
fr structure polaire-apolaire
Structure d'une molécule possédant au moins un groupement polaire et un radical apolaire important. Une telle structure conditionne les caractères d'hydrophilie et de lipophilie de la molécule.
de polar-apolare Struktur
Struktur eines Moleküls, die durch das Vorhandensein wenigstens einer polaren Gruppe und eines grösseren apolaren Restes bedingt ist. Eine solche Struktur bedingt den hydrophilen und den lipophilen Charakter des Moleküls.
el estructura polar-apolar
Estructura de una molécula que posee al menos un grupo polar y un radical apolar importante. Esta estructura confiere a la molécula sus carácteres de hidrofilia y de lipofilia.
it struttura polare-apolare
Struttura della molecola avente almeno un gruppo polare ed un radicale apolare importante. Tale struttura determina i caratteri di idrofilia e lipofilia della molecola.
ne polaire/niet-polaire struktuur
De struktuur van een molecuul met ten minste één polaire en een grote niet-polaire groep. Een dergelijke struktuur bepaalt het hydrofiele en lipofiele karakter van het molecuul.
pl budowa polarno-niepolarna
Budowa cząsteczki, w której występuje co najmniej jedna grupa polarna oraz duża grupa niepolarna. Budowa taka warunkuje hydrofilowy i lipofilowy charakter cząsteczki.

458 polish remover
fr dissolvant
de Nagellackentferner
el disolvente
it solvente
ne nagellakverwijderer
pl zmywacz do lakieru do paznokci

459 polishing agent
fr agent de polissage
de Poliermittel
el agente pulimentador
it agente di pulitura
ne polijstmiddel
pl środek polerujący

460 polypeptide chain
fr chaîne polypeptidique
de Polypeptidkette
el cadena polipéptida
it catena polipeptidica
ne polypeptide-keten
pl łańcuch polipeptydowy

461 potash soft soap
fr savon mou à la potasse
de Kalischmierseife
el jabón blando de potasa, jabón blando potásico
it sapone molle di potassa
ne zachte kalizeep
pl mydło maziste potasowe

462 potassium bromate
fr bromate de potasse, bromate de potassium
de Kaliumbromat
el bromato potásico
it bromato di potassio
ne kaliumbromaat
pl bromian potasu

463 powder base
fr base de poudre
de Pudergrundlage
el base de polvos
it cipria di base
ne poederbasis
pl baza pudrowa

464 powder stick
fr poudre en bâton
de Puderstift
el polvo en barra
it polvere "stick"
ne poederstift
pl puder w sztyfcie, puder w postaci pałeczki

465 powdered shaving soap
fr savon à barbe en poudre

de pulverförmige Rasierseife
el jabón de afeitar en polvo
it sapone da barba in polvere
ne poedervormige scheerzeep
pl mydło do golenia w postaci proszku

466 powdered soap
fr savon en poudre
de Seifenpulver
el jabón en polvo
it sapone in polvere
ne zeeppoeder
pl proszek mydlany

467 preparing agent, pre-processing agent
Product intended, in general, to make a textile material better suited to undergo a subsequent operation, as spinning, winding, knitting, etc., by reducing, for example, the friction. Some of these products are also brightening agents in finishing processes.
Note: These products are generally surface active agents or preparations of the latter with oils and greases. The surface active agents used include, for example, sulphated oils and greases, alkylsulphates, esters and amides of fatty acids, fatty amine condensates and also the ethoxylation products of fatty acids and fatty alcohols, fatty amides or fatty amine condensates.

fr agent de préparation
Produit destiné, d'une façon générale, à rendre une matière textile mieux adaptée à subir une opération subséquente, comme filature, bobinage, tricotage, etc., en diminuant par exemple, le frottement. Certains de ces produits sont également utilisés comme agent d'avivage dans les traitements de finition.
Nota: Il s'agit en général d'agents de surface ou de préparations de ces derniers avec des huiles et des graisses. Les agents de surface utilisés sont par exemple: huiles et graisses sulfatées, alkylsulfates, esters et amides d'acides gras, condensats d'amines grasses, ainsi que les produits d'éthoxylation d'acides gras et d'alcools gras, amides gras ou condensats d'amines grasses.

de Präparationsmittel
Produkt, das generell betrachtet, die Weiterverarbeitung von Textilmaterialien in den nachfolgenden Prozessen, wie Strecken, Spinnen, Weben, Wirken, beispielsweise durch Verminderung der Reibung, ermöglicht oder erleichtert. Einige dieser Produkte werden auch als Avivagemittel in der Endausrüstung verwandt.
Anmerkung: Es handelt sich in der Regel um grenzflächenaktive Stoffe oder Zubereitungen hieraus mit Ölen und Fetten. Als grenzflächenaktive Stoffe werden beispielsweise eingesetzt: sulfierte Öle und Fette, Alkylsulfate, Fettsäureester und -amide, Fettsäureaminkondensationsprodukte, sowie Äthoxylierungsprodukte von Fettsäuren und Fettalkoholen, Alkylaminen, Fettsäureamiden oder Fettsäureaminkondensationsprodukten.

el agente de preparación
Producto que, en general, sirve para preparar una materia textil para los tratamientos posteriores, como estiraje, hilatura, tisaje, tricotaje, etc. Algunos de estos productos se utilizan igualmente como agentes de avivado en los tratamientos de acabado.
Observación: Se trata en general, de agentes de superficie o de sus preparaciones con aceites y grasas. Los agentes de superficie utilizados son: aceites y grasas sulfatados, alquilsulfatos, ésteres y amidas de ácidos grasos, productos de condensación de aminas grasas, así como los productos de oxietilación de ácidos grasos y de alcoholes grasos, de alquilaminas, de amidas grasas o de productos de condensación de aminas grasas.

it —
ne avivage
Een produkt dat over het algemeen bestemd is om de produktie van synthetische vezels mogelijk te maken, vezels, garen of doek beter geschikt te maken voor een volgende bewerking of tot doel heeft het materiaal speciale eigenschappen te verlenen.

pl środek do preparacji
Produkt przeznaczony, ogólnie biorąc, do lepszego przystosowania wyrobu włókienniczego do takich operacji jak przędzenie, przewijanie, dzianie, tkanie itd. przez zmniejszenia np. tarcia. Niektóre z tych produktów są również stosowane jako środki ożywiające w procesach wykańczania.
Uwaga: Ogólnie biorąc produktami takimi są związki powierzchniowo czynne lub zawierające je preparaty otrzymane z zastosowaniem olejów lub tłuszczów. Stosowane związki powierzchniowo czynne obejmują np. siarczanowane oleje i tłuszcze, alkilosiarczany, estry i amidy kwasów tłuszczowych, produkty kondensacji amin tłuszczowych, oraz produkty oksyetylenowania kwasów, alkoholi i amidów kwasowych oraz produktów kondensacji amin tłuszczowych.

468 pre-shaving lotion, pre-shaving preparation
fr lotion avant rasage, préparation avant rasage
de Rasiermittel zur Vorbehandlung
it lozione pre rasatura
el loción para antes del afeitado, preparado para antes del afeitado
ne gezichtswater voor vóór het scheren
pl płyn przed goleniem

469 pressure atomisation
fr pulvérisation sous pression
de Düsenzerstäubung
el pulverización a presión
it atomizzazione sotto pressione
ne verstuiving onder druk
pl rozpylanie ciśnieniowe

470 pressure pack
fr emballage sous pression
de Druckpackung
el envase a presión
it imballaggio sotto pressione
ne drukverpakking
pl opakowanie ciśnieniowe, opakowanie aerozolowe

471 propoxylation
In the particular case of surface active agents, chemical reaction leading to the addition of one or more molecules of propylene oxide to a labile hydrogen compound.
fr propoxylation (propoxylénation)
Dans le cas particulier des agents de surface, réaction chimique permettant la fixation d'une ou plusieurs molécules d'oxyde de propylène sur un composé à hydrogène labile.
de Propoxylierung
Im speziellen Fall der grenzflächenaktiven Körper: chemische Reaktion, bei welcher ein oder mehrere Mol Propylenoxyd an eine Verbindung mit aktivem Wasserstoff angelagert werden.
el propoxilación (propoxilenación)
En el caso particular de los agentes de superficie: reacción química, que permite la fijación de una o de varias moléculas de óxido de propileno, en un compuesto con hidrógeno lábil.
it propossilazione
Nel caso particolare dei tensioattivi: reazione chimica di addizione di una o più molecole di ossido di propilene su un composto ad idrogeno mobile.
ne propoxylering
In het bijzondere geval van oppervlakaktieve stoffen is dit een chemische reactie waarbij een of meer moleculen van propyleenoxide geaddeerd worden aan een labiele waterstofverbinding.

pl propoksylowanie
W szczególnym przypadku związków powierzchniowo czynnych reakcja chemiczna prowadząca do przyłączenia jednej lub więcej cząsteczek tlenku propylenu do związku posiadającego labilny wodór.

472 protective colloid
Substance which, within a certain concentration range and when acting as a lyophilic colloid, retards or prevents the aggregation of the particles of a lyophobic dispersion.
fr colloïde protecteur
Substance qui, à une certaine gamme de concentrations et agissant en l'état de colloïde lyophile, retarde ou empêche l'agrégation des particules d'une dispersion lyophobe.
de Schutzkolloid
Lyophiles Kolloid, welches innerhalb bestimmter Konzentrationsbereiche die Aggregation der Partikel einer hydrophoben Dispersion verzögert oder verhindert.
el coloide protector
Substancia que, a una cierta gama de concentraciones y actuando como coloide liófilo, retrasa o impide la agregación de las partículas de una dispersión liófoba.
it colloide protettore
ne schutkolloid
Lyophiel kolloid, dat tussen bepaalde grenzen van zijn concentratie het samenstellen van deeltjes van een hydrophobe dispersie vertraagt of verhindert.
pl koloid ochronny
Substancja, która w pewnym zakresie stężenia działając jako koloid liofilowy, opóźnia lub zapobiega agregacji cząstek dyspersji liofobowej.

473 purifying mouthwash
fr bain de bouche purifiant
de reinigendes Mundwasser
el lavado de boca purificador
it lavaggio purificante della bocca
ne reinigend mondspoelmiddel
pl oczyszczająca woda do płukania ust

Q R

474 **quantity of soil**
fr quantité de la souillure
de Schmutzmenge
el cantidad de suciedad
it quantità dello sporco
ne hoeveelheid vuil
pl ilość brudu, ilość zabrudzenia

475 **rancidity**
fr rancissement, rancidité
de Ranzigwerden, Ranzidität
el enranciamiento
it irrancidimento, rancidità
ne rans worden
pl jełczenie

476 **razor blade**
fr lame de rasoir
de Rasierklinge
el hoja de afeitar
it lametta da barba
ne scheermesje
pl żyletka

477 **receding wetting angle**
The withdrawal of a solid surface at slow and constant speed from a liquid phase gives rise to the formation of a contact angle the dimension of which may depend upon the nature of the surface and the speed of withdrawal. This angle is called the receding wetting angle.

Note: 1. As generally measured, the receding wetting angle is that which corresponds to withdrawal perpendicular to the surface of the liquid phase.
2. In the case of a solid surface, the advancing and receding wetting angles relating to a liquid phase are different. The difference arises from measurement of the angles produced at the speed of introduction and withdrawal laid down in the test method.

fr angle de mouillage sortant
L'extraction lente et à vitesse constante d'une surface solide vers l'extérieur d'une phase liquide donne lieu à la formation d'un angle de contact dont la grandeur peut dépendre de la nature de la surface et de la vitesse d'extraction. Cet angle est appelé angle de mouillage sortant.

Nota: 1. Un angle de mouillage sortant, généralement mesuré, est celui qui correspond à une extraction effectuée perpendiculairement à la surface de la phase liquide.
2. Dans le cas d'une surface solide, les angles de mouillage rentrant et sortant vis-à-vis d'une phase liquide sont différents. La différence résulte de mesures d'angles effectuées dans des domaines de vitesse de pénétration et d'extraction, précisées par les méthodes d'essais.

de rückläufiger Randwinkel
Das langsame und mit konstanter Geschwindigkeit erfolgende Herausziehen einer Festkörperoberfläche aus einer flüssigen Phase bedingt die Bildung eines Kontaktwinkels, dessen Grösse von der Natur der Oberfläche und von der Extrak-

tionsgeschwindigkeit abhängen kann. Dieser Winkel wird als rückläufiger Randwinkel bezeichnet.

Anmerkungen: 1. Ein im allgemeinen gemessener ruckläufiger Randwinkel entspricht einem Herausziehen senkrecht zur Oberfläche der flüssigen Phase.
2. Im Falle einer festen Oberfläche sind die fortschreitenden und rückläufigen Randwinkel verschieden. Die Differenz resultiert aus Messungen der Winkel in einem Geschwindigkeitsgebiet des Eintauchens und Herausziehens, die durch Testmethoden präzisiert werden.

el ángulo de mojado saliente
La extracción lenta y a velocidad constante de una superficie sólida hacia el exterior de una fase líquida da lugar a la formación de un ángulo de contacto, cuyo valor puede depender de la naturaleza de la superficie y de la velocidad de extracción. Este ángulo se llama ángulo de mojado saliente.

Observaciones: 1. El ángulo de mojado saliente, el mas comúnmente medido, corresponde a una extracción efectuada perpendicularmente a la superficie de la fase líquida.
2. En el caso de una superficie sólida, los ángulos de mojado entrante y saliente, con respecto a una fase líquida, son diferentes. La diferencia resulta de medidas efectuadas a las velocidades de penetración o de extracción prescritas en los métodos de ensayo.

it angolo di bagnatura regressiva
L'estrazione di una superficie solida da una fase liquida, a velocità lenta e costante dà luogo alla formazione di un angolo di contatto che può dipendere dalla natura della superficie e dalla velocità di estrazione. Tale angolo è chiamato angolo di bagnatura regressiva.

Note: 1. L'angolo di bagnatura regressiva, come viene generalmente misurato, è quello che corrisponde ad una estrazione effettuata perpendicolarmente alla superficie della fase liquida.
2. Per una superficie solida l'angolo di bagnatura progressiva e l'angolo di bagnatura regressiva, relativi ad una fase liquida, sono differenti. La differenza risulta dalle misure degli angoli effettuate alle velocità di introduzione e di estrazione precisate nei metodi di prova. Il lavoro necessario per introdurre lentamente e ad una velocità costante una parte di una superficie solida in una fase liquida è, in certe condizioni, diverso da quello corrispondente all'estrazione, nelle stesse condizioni, della stessa parte di superficie dalla stessa fase liquida. La differenza fra questi due lavori, riferita all'unità di superficie, rappresenta l'isteresi di bagnatura; si esprime in erg ed è numericamente uguale alla differenza fra la tensione di bagnatura progressiva e la tensione di bagnatura regressiva.

ne regressierandhoek
Wanneer een vast oppervlak langzaam en met konstante snelheid uit een vloeibare fase wordt getrokken, ontstaat een randhoek afhankelijk van de aard van het oppervlak en de snelheid van het uittrekken. Deze hoek wordt de "regressierandhoek" genoemd.

Opmerkingen: 1. Gewoonlijk wordt de regressierandhoek gemeten bij vertikaal uit het vloeistofoppervlak trekken.
2. In het geval van een vast oppervlak is er verschil tussen de progressie- en regressierandhoek ten opzichte van een vloeibare fase. Het verschil ontstaat door meting van de hoeken die ontstaan bij de snelheid van inbrengen en uitnemen welke is vastgelegd in de beproevingsmethode.

pl zstępujący kąt zwilżania
Wyjmowanie powolne i ze stałą prędkością powierzchni ciała stałego z fazy ciekłej powoduje powstanie kąta zwilżania, którego wartość może zależeć od charakteru powierzchni i od szybkości wyjmowania. Kąt ten nazywany jest zstępującym kątem zwilżania.

Uwagi: 1. Normalnie mierzony zstępujący kąt zwilżania występuje przy wyjmowaniu ciała stałego prostopadle do powierzchni fazy ciekłej.
2. Dla danej powierzchni ciała stałego występuje różnica pomiędzy wstępującym i zstępującym kątem zwilżania. Różnica ta wynika z pomiarów kątów powstałych przy zanurzaniu i wyjmowaniu z szybkością podaną w metodzie badań.

478 **receding wetting tension**
Wetting tension corresponding to the formation of a receding wetting angle.

fr tension de mouillage sortante
Tension de mouillage correspondant à la formation d'un angle de mouillage sortant.

de rückläufige Benetzungsspannung
Ist die einem rückläufigen Randwinkel entsprechende Benetzungsspannung.

el tensión de mojado saliente
Tensión de mojado correspondiente a la formación de un ángulo de mojado saliente.

it tensione di bagnatura regressiva
Tensione di bagnatura corrispondente alla formazione di un angolo di bagnatura regressiva.

ne regressiebevochtigingsspanning
Bevochtigingsspanning die overeenkomt met de vorming van een regressiebevochtigingshoek.

pl zstępujące napięcie zwilżania
Napięcie zwilżania odpowiadające powstawaniu zstępującego kąta zwilżania.

479 **red stripes**
fr bandes rouges
de rote Streifen
el rayas rojas
it strisce rosse

ne rode strepen
pl czerwone paski

480 redness
fr rougeur
de Röte
el enrojecimiento
it arrossamento
ne roodheid
pl rumień

481 reduction
fr réduction
de Reduktion
el reducción
it riduzione
ne reductie
pl redukcja

482 reduction inhibitor
Product lessening the reducing effect of foreign matter on dyes, and consequently combating the destruction of the latter.
Note: The products include, for example, preparations based on buffer substances and oxidizing substances with surface active agents, such as: degraded proteins, fatty acid and protein condensates, ammonium salts of alkylsulphates and alkylsulphonates.

fr agent antiréducteur
Produit diminuant l'action réductrice de matières étrangères sur des colorants et, par conséquent, s'opposant à la destruction de ces derniers.
Nota: Il s'agit, par exemple, de préparations à base de substances tampons et de substances oxydantes avec des agents de surface, telles que protéines dégradées, condensats d'acides gras et de protéines, sels ammoniacaux d'alkylsulfates et d'alkylsulfonates.

de Mittel gegen das Farbstoffverkochen
Produkt, das die reduzierende Wirkung von Fremdsubstanzen auf Farbstoffe und damit ihre Zerstörung vermindert.
Anmerkung: Es handelt sich z.B. um Zubereitungen aus puffernd und oxydierend wirkenden Substanzen mit grenzflächenaktiven Stoffen, wie Eiweissspaltprodukte, Fettsäure-Eiweisskondensationsprodukte, Ammoniumsalze von Alkylsulfaten und Alkylsulfonaten.

el agente anti-reductor
Producto que disminuye la acción reductora de substancias extrañas sobre los colorantes, oponiéndose así a la destrucción de éstos.
Observación: Se trata, por ejemplo, de preparaciones de agentes de superficie con substancias de acción reguladora y oxidantes, tales como proteínas degradadas, productos de condensación de ácidos grasos y proteínas, así como sales amoniacales de alquilsulfatos y de alquilsulfonatos.

it —
ne reductievertragend middel

Een produkt dat een ongewenst reducerend effekt van bepaalde chemicaliën op een verving tegengaat.
pl inhibitor redukcji
Produkt zmniejszający redukujące działanie zanieczyszczeń na barwniki i przeciwdziałający w rezultacie ich rozkładowi.
Uwaga: Produkty te obejmują np. preparaty oparte na substancjach buforujących i substancjach utleniających, zawierające takie związki powierzchniowo czynne jak produkty degradacji białek, produkty kondensacji białek z kwasami tłuszczowymi oraz sole amonowe alkilosiarczanów i alkilosulfonianów.

483 reduction in surface tension, lowering of the surface tension
fr abaissement de la tension superficielle
de Oberflächenspannungserniedrigung
el reducción de la tensión superficial, disminución de la tensión superficial
it abbassamento di tensione superficiale
ne verlaging van de oppervlaktespanning
pl obniżenie napięcia powierzchniowego

484 reflectance measurement, reflectance reading
fr mesure de réflexion, mesure de réflectance
de Reflexionsmessung
el medida de reflexión
it misura della riflessione
ne reflectie-meting
pl pomiar współczynnika odbicia światła

485 refreshing lotion
fr lotion rafraîchissante
de Erfrischungsparfüm
el loción refrescante
it lozione rinfrescante
ne verfrissend toiletwater
pl płyn odświeżający

486 replace cap firmly
fr replacer le bouchon soigneusement, revisser à fond
de Kappe wieder fest aufsetzen, Deckel wieder fest aufsetzen
el volver a colocar el tapón firmamente, cerrar bien
it richiudere bene
ne sluit de dop krachtig
pl po użyciu szczelnie zamknąć

487 residual hardness
fr dureté permanente
de bleibende Härte
el dureza permanente
it durezza permanente

ne blijvende hardheid
pl twardość trwała

488 residual lime soap
fr savon de chaux résiduaire, savon de chaux restant
de Restkalkseife
el jabón de cal residual
it sapone di calcio residuo
ne rest-kalkzeep
pl pozostałość mydła wapniowego

489 resistance at the boil
fr résistance à l'ébullition
de Kochbeständigkeit
el resistencia a la ebullición
it resistenza all'ebollizione
ne kookbestendigheid
pl odporność na gotowanie

490 reversible hydrolysis
The action of water on ions from a dissolved salt, which assumes a state of equilibrium in which exist both ions and molecules of acid or base capable of forming the salt. The molecules of acid or base can revert to the ionic state when the conditions in the medium change. Reversible hydrolysis is particularly noticeable in the case of salts of weak organic acids or weak organic bases, possessing appreciable hydrophobic radicals.

fr hydrolyse réversible
Action de l'eau sur les ions correspondant à un sel dissous, qui tend à l'établissement d'un état d'équilibre où coexistent les ions et des molécules d'acide ou de base assurant la formation du sel. Les molécules d'acide ou de base peuvent inversement repasser à l'état d'ions, quand les conditions du milieu varient. L'hydrolyse réversible s'observe particulièrement pour les sels d'acides organiques faibles ou d'amines organiques faibles, à radicaux hydrophobes importants.

de reversible Hydrolyse
Wirkung des Wassers auf die Ionen eines gelösten Salzes, die zu einem Gleichgewichtszustand führt, bei dem die Ionen und die Moleküle der Säure oder der Base, die das Salz bilden, nebeneinander vorhanden sind. Die Säure- oder Basenmoleküle können umgekehrt wieder in Ionen übergehen, wenn sich die Bedingungen des Milieus verändern. Die reversible Hydrolyse kommt speziell vor bei Salzen von schwachen organischen Säuren oder schwachen organischen Aminen mit grossen hydrophoben Resten.

el hidrólisis reversible
Acción del agua sobre los iones de una sal disuelta, que conduce a un estado de equilibrio, en el que coexisten los iones y moléculas de ácido o de base que forma la sal. Las moléculas de ácido o de base pueden pasar inversamente al estado de iones cuando las condiciones del medio varían. La hidrólisis reversible se observa en especial en las sales de ácidos orgánicos débiles o de bases orgánicas débiles, con radicales apreciablemente hidrófobos.

it idrolisi reversibile
Azione dell'acqua sugli ioni di un sale disciolto che tende a stabilire uno stato di equilibrio nel quale coesistano gli ioni e le molecole dell'acido o della base, che formano il sale. Le molecole dell'acido e della base possono ritornare allo stato di ioni al variare delle condizioni del mezzo. L'idrolisi reversibile si osserva in particolare per i sali di acidi organici deboli o di ammine organiche deboli, con radicali fortemente idrofobi.

ne omkeerbare hydrolyse
De werking van water op de ionen van een oplosbaar zout die een evenwichtstoestand tot stand brengt, waarbij zowel ionen als moleculen van het zuur of de base die het zout kunnen vormen aanwezig zijn. De moleculen van het zuur of de base kunnen in ionvorm terugkeren als de omstandigheden van het medium veranderen. Omkeerbare hydrolyse wordt vooral waargenomen bij zouten van zwakke organische zuren of zwakke organische basen die sterk hydrofobe radicalen bezitten.

pl hydroliza odwracalna
Oddziaływanie wody na jony rozpuszczonej soli, prowadzące do stanu równowagi, w którym występują obok siebie zarówno jony jak i cząsteczki kwasu lub zasady tworzącej sól. Cząsteczki kwasu lub zasady mogą powracać do stanu jonowego przy zmianie warunków środowiska. Hydroliza odwracalna występuje w szczególności w przypadku

soli słabych kwasów organicznych lub
słabych zasad organicznych zawierających duże rodniki hydrofobowe.

491 rheopexy
Under isothermal and reversible conditions, increase with hysteresis of the
apparent viscosity under a shearing load.
fr rhéopexie, anti-thixotropie
Dans des conditions isothermes et réversibles, augmentation avec hystérésis
de la viscosité apparente sous l'effet
d'une contrainte de cisaillement croissante.
de Rheopexie
Unter isothermen und reversiblen Bedingungen, Vergrösserung (mit Hysterese)
der scheinbaren Viskosität unter dem
Einfluss einer mechanischen Schubspannung.
el reopexia
En condiciones isotermas y reversibles,
aumento con histéresis de la viscosidad
aparente, por la acción de un esfuerzo
de cizallamiento creciente.
it reopexia
ne reopexie
Onder isotherme en reversibele omstandigheden, verhoging met hysterese van de
schijnbare viscositeit onder de invloed
van een mechanische afschuifspanning.
pl reopeksja
Wzrost (z histerezą) lepkości pozornej
pod wpływem działania obciążenia ścina-
jącego w warunkach izotermicznych
i odwracalnych.

492 rinsing
fr rinçage
de Spülen
el enjuague
it risciacquo
ne spoelen
pl płukanie

493 rinsing agent
fr produit de rinçage, agent à rincer
de Spülmittel
el producto del aclarado, agente del aclarado

it agente di risciacquatura
ne spoelmiddel
pl środek do płukania

494 rinsing solution
fr eau de rinçage, bain de rinçage
de Spülwasser, Spülbad
el agua de aclarar
it acqua di risciacquatura
ne spoelwater
pl kąpiel płucząca

495 road dust (natural)
fr poussière naturelle de la rue
de Strassenstaub
el polvo de la calle
it polvere (naturale) di strada
ne straatvuil
pl kurz uliczny (naturalny)

496 roller (for the roller wave)
fr rouleau (pour la mise en plis)
de Wickler (für die Wickelwelle)
el tubos (para marcar)
it rulli (per la messa in piega)
ne watergolfpen
pl wałek (do nawijania włosów)

497 roller press
fr broyeuse
de Walzenpresse
el desmenuzadora
it truciolatrice
ne wals
pl prasa walcowa

498 rouge
fr rouge gras
de Rouge
el rojo graso
it rossetto
ne rode schmink
pl róż

499 rubefacient
fr rubéfiant
de hautrötendes Mittel
el colorete
it belletto
ne roodmakend middel
pl środek wywołujący rumień

S

500 saliva, spittle
fr salive
de Speichel
el saliva
it saliva
ne speeksel
pl ślina, plwocina

501 salivary gland
fr glande salivaire
de Speicheldrüse
el glándula salivar
it ghiandola salivaria
ne speekselklier
pl gruczoł ślinowy

502 salivation
fr salivation, ptyalisme
de Speichelfluss
el flujo salivar
it flusso salivare
ne speekselafscheiding
pl ślinienie, ślinotok

503 salt bridge
fr pont salin
de Salzbrücke
el puente salino
it ponte salino
ne zoutbrug
pl mostek elektrolityczny, klucz elektrolityczny, mostek jonowy

504 salt linkage
fr liaison saline
de Salzbindung
el enlace salino
it legame salino
ne zoutband
pl wiązanie jonowe

505 saponification (in the soap industry)†
A chemical reaction which converts
a fatty body into soap by reaction with
a base. In the case of a fatty acid, this
term is sometimes used to imply a simple
neutralisation. In the saponification of
a neutral fat, glycerine is produced at
the same time.
fr saponification (dans l'industrie des savons)†
Réaction chimique de transformation
d'un corps gras en savon par l'action
d'une base. Dans le cas d'un acide gras,
la réaction est une simple neutralisa-

tion. Dans le cas d'un corps gras, il se
forme en même temps de la glycérine.
de Verseifung (in der Seifenindustrie)†
Chemische Reaktion der Umsetzung von
Fettstoffen in Seife durch Einwirkung
einer Base. Bei Fettsäuren ist die Reaktion, die einer einfachen Neutralisation.
In der Verseifung von Neutralfett wird
gleichzeitig Glycerin frei.
el saponificación
En la industria de jabones se llama
saponificación la reacción química de
transformación de grasas y/o ácidos
grasos por la acción de substancias alcalinas adecuadas. En la saponificación
de grasas se forma glicerina al mismo
tiempo.
it saponificazione
Reazione chimica che consente di sepa-
rare gli elementi costitutivi di un estere:
acido ed alcole (eventualmente fenolo)
per mezzo di una base che forma un
sale con l'acido. La saponificazione dei
grassi porta alla formazione di sapone.
ne verzeping (in de zeepindustrie)
Chemische reactie van de omzetting van
vetten in zeep door inwerking van een
base. Bij vetzuren is de reactie die van
een eenvoudige neutralisatie. Bij het
verzepen van een neutraal vet komt
gelijktijdig glycerine vrij.
pl zmydlanie (w przemyśle mydlarskim)
Reakcja chemiczna, w której substancja
tłuszczowa zostaje przeprowadzona
w mydło pod działaniem zasady. W przypadku kwasów tłuszczowych jest to
zwykła reakcja zobojętnienia. Przy zmy-
dlaniu tłuszczów obojętnych równocze-
śnie z mydłem powstaje gliceryna.
† See appendix
  Voir appendice
  Siehe Appendix

506 scalp
fr cuir chevelu
de Kopfhaut
el cuero cabelludo
it cuoio capelluto
ne hoofdhuid
pl owłosiona skóra głowy

507 sebacceous-gland secretion (excretion)
fr sécrétion sébacée, excrétion sébacée

de Talgdrüsenabsonderung, Talgdrüsen-
ausscheidung
el secreción sebácea
it secrezione sebacea, escrezione sebacea
ne afscheiding der vetklieren
pl wydzielina gruczołów łojowych, wydzielanie gruczołów łojowych

508 sedimentation
In a liquid medium, the accumulation
of particles in dispersion under the
effect of gravity or centrifugal force.
fr sédimentation
Dans un milieu fluide, accumulation de
particules en dispersion, sous l'effet
d'une force de gravité ou d'une force
centrifuge.
de Sedimentation
Anreicherung der dispergierten Teilchen
in einem flüssigen Medium unter der
Einwirkung der Schwerkraft oder einer
Zentrifugalkraft.
el sedimentación
En un medio fluido, acumulación de
partículas en dispersión, bajo la acción
de la gravedad o de una fuerza centrí-
fuga.
it sedimentazione
ne bezinking
Verrijking van de gedispergeerde deeltjes
in een vloeibaar medium onder de
inwerking van de zwaartekracht of een
centrifugale kracht.
pl sedymentacja, osadzanie się
Oddzielanie się cząstek zdyspergowa-
nych od cieczy pod wpływem siły cięż-
kości lub siły odśrodkowej.

509 semi-boiled soap
Soap prepared without graining out, by
saponification at the boil of fatty matter
with a quantity of alkali hydroxide just
sufficient to complete the reaction. In
the case of fats, the glycerine formed
remains in the mass of the soap.
fr savon mi-cuit
Savon préparé sans relargage par saponification à l'ébullition des corps gras,
avec la quantité d'hydroxyde alcalin
juste suffisante pour compléter la réaction. Dans le cas de corps gras, la glycérine formée reste dans la masse du
savon.
de Leimseife
Seife ohne Elektrolytüberschuss, die
durch Verseifung ohne Aussalzen oder

Ausschleifen mit der berechneten Alkalimenge hergestellt wird. Bei Verwendung
von Neutralfett bleibt das gebildete
Glycerin in der Seife zurück.
el jabón de empaste en caliente
Jabón preparado sin exceso de electrólitos, por saponificación, sin saladura,
con la cantidad calculada de hidróxido
sódico. Si se emplean grasas neutras,
la glicerina formada queda en la masa
de jabón.
it sapone d'impasto (sapone semicotto)
Il sapone preparato senza salatura per
saponificazione alla ebollizione di corpi
grassi, con il quantitativo di alcali
caustico, strettamente necessario per
completare la reazione. Nel caso dei
grassi, la glicerina che si produce, rimane
nella massa del sapone.
ne lijmzeep
Zeep zonder overmaat aan electrolyt,
die door verzeping zonder uitzouten
of afmaken met de berekende hoeveelheid alkali bereid wordt. Bij gebruik
van neutraal vet blijft het gevormde
glycerine in de zeep achter.
pl mydło klejowe
Mydło otrzymane bez operacji wysolenia
przez zmydlenie substancji tłuszczowej
przy użyciu takiej ilości alkaliów, która
wystarcza dla zakończenia reakcji.
W przypadku tłuszczów obojętnych
powstała gliceryna pozostaje w mydle.

510 sensitising
fr sensibilisation
de Sensibilisierung
el sensibilización
it sensibilizzazione
ne sensibilisatie
pl uczulanie

511 sequestering agent, sequestrant
A substance having functional characteristics which make it capable of sup-
pressing metallic ions and ensuring that
they remain in solution in the medium.
fr agent séquestrant, séquestrant
Produit possédant des caractéristiques
fonctionnelles le rendant apte à la fois
à dissimuler les ions métalliques et
à assurer leur solubilisation dans le
milieu.
de Komplexbildner
Produkt, das durch charakteristische
funktionelle Gruppen befähigt ist, Me-

tallionen durch Komplexbildung zu verändern und ihre Löslichkeit in einem bestimmten Milieu zu verbessern.

el agente secuestrante, secuestrante
Producto que posee grupos funcionales característicos, capaces de enmascarar iones metálicos y de asegurar su solubilización en el medio.

it sequestrante
Sostanza con caratteristiche funzionali che la rendono idonea a complessare gli ioni metallici assicurandone la solubilizzazione nel mezzo.

ne sekwestreermiddel
Een stof met funktionele kenmerken die de ionisatie van het metaalion terugdringt en deze in het medium in oplossing houdt.

pl środek sekwestrujący, środek kompleksujący
Substancja, która dzięki charakterystycznym grupom funkcyjnym zdolna jest do usuwania jonów metalicznych i utrzymania w roztworze.

512 sequestering power
The ability of certain substances to keep cations in solution, in a more or less labile condition, so that the reactions of the cations are then, for the most part, masked.

fr pouvoir séquestrant
Aptitude de certains corps à retenir (en solution), d'une manière plus ou moins labile, des cations dont les réactions sont alors généralement dissimulées.

de Sequestriervermögen
Fähigkeit gewisser Körper, Kationen mehr oder weniger stabil in Lösung zu halten, wobei die chemischen Reaktionen der Kationen im allgemeinen verändert werden.

el poder secuestrante
Capacidad de ciertos cuerpos para retener en disolución, en forma más o menos lábil, cationes cuyas reacciones quedan generalmente enmascaradas.

it potere sequestrante
Grado dell'attitudine di certe sostanze a mantenere in soluzione, in modo più o meno labile, dei cationi le cui reazioni risultano generalmente inibite.

ne sekwestreervermogen
Het vermogen van bepaalde stoffen kationen in een min of meer labiele

toestand in oplossing te houden, zodat de reacties van de kationen dan grotendeels gemaskeerd worden.

pl zdolność sekwestrująca
Zdolność pewnych substancji do utrzymywania kationów w roztworze w sposób bardziej lub mniej trwały tak, że w większości przypadków reakcje chemiczne tych kationów są maskowane.

513 sequestration
The masking of metallic ions dissolved in a medium, the ions being normally liable to form precipitates in the presence of certain reagents, particularly surface active agents. The masking is accomplished by the formation of complexes which remain in solution in the medium.

fr séquestration
Dissimulation d'ions métalliques dissous dans un milieu, susceptibles de former des précipités en présence de certains réactifs et d'agents de surface en particulier. La dissimulation s'effectue par formation de complexes restant nécessairement solubles dans le milieu.

de Sequestrierung
Umwandlung eines in einer flüssigen Phase gelösten Metallions, das mit gewissen Reagenzien, insbesondere mit grenzflächenaktiven Verbindungen, Niederschläge bilden kann. Diese Umwandlung bewirkt die Bildung eines Metallkomplexes, der mit den betreffenden Reagenzien in der flüssigen Phase löslich bleibt.

el secuestración
Transformación estructural de iones metálicos, disueltos en un medio, que son capaces de formar precipitados con ciertos reactivos y en particular con agentes de superficie. Esta transformación se efectúa por formación de complejos que, aun en presencia de los referidos reactivos, quedan disueltos en el medio.

it sequestrazione
Dissimulazione di ioni metallici disciolti in un mezzo ed in grado di dare precipitati in presenza di determinati reattivi e di tensioattivi in particolare. I complessi formati devono restare necessariamente disciolti nel mezzo.

ne sekwestratie
Het maskeren van metaalionen, opgelost

in een medium waarin de ionen nor-
maliter een neerslag zouden vormen in
aanwezigheid van bepaalde reagentia,
in het bijzonder oppervlakaktieve stoffen.
Het maskeren wordt bereikt door de
vorming van complexen die in het
medium op natuurlijke wijze in oplossing
blijven.

pl sekwestracja
Przemiana rozpuszczonych w środowisku
ciekłym jonów metali, które normalnie
w obecności pewnych związków chemicznych (zwłaszcza związków powierzchniowo czynnych) tworzą osady,
w związki kompleksowe pozostające
w roztworze.

514 setting lotion
fr lotion de mise en plis
de festigende Lösung
el loción para marcar
it lozione per la messa in piega
ne watergolf
pl płyn do układania włosów

515 shade, hue
fr nuance
de Farbton, Farbstich
el matiz
it nuanza
ne schakering, nuance
pl odcień

516 shaped washing agent
fr produit de lavage moulé, produit de
lavage enformé
de geformtes Waschmittel
el detergente compuesto
it detergente in pezzi
ne gevormd wasmiddel
pl środek piorący w kawałku

517 shaving brush
fr blaireau
de Rasierpinsel
el brocha
it pennello da barba
ne scheerkwast
pl pędzel do golenia

518 shaving cream
fr crème à raser
de Rasierseifencreme
el crema de afeitar
it crema da barba
ne scheercrème
pl krem do golenia

519 shaving soap
Soap obtained from fats which contain
a high percentage of stearic acid, saponified with a mixture of sodium and
potassium hydroxides.
fr savon à raser
Savon obtenu avec un mélange de corps
gras contenant une proportion élevée
d'acide stéarique, saponifié avec des
hydroxydes de sodium et de potassium.
de Rasierseife
Mit Kali- und Natronlauge verseifte
Seifen aus Fettsätzen, die vergleichs-
weise hohe Anteile an Stearinsäure
enthalten.
el jabón de afeitar
Jabón con 76 a 81% de ácidos grasos,
saponificados con lejías de potasa y sosa,
que contienen una proporción relativa-
mente elevada de estearina.
it sapone da barba
Saponi ottenuti da grassi ad alta percen-
tuale di acido stearico, saponificati con
miscele di potassio e sodio idrato.
ne scheerzeep
Met kali- en natronloog verzeepte zepen,
die verhoudingsgewijs een hoog gehalte
aan stearinezuur hebben.
pl mydło do golenia
Mydło otrzymane z tłuszczów o dużej
zawartości kwasu stearynowego, zmydlo-
nych za pomocą mieszaniny wodorotlenków sodowego i potasowego.

520 shaving stick
fr bâton à raser, savon à barbe
de Rasierseifenstange
el barra de jabón de afeitar
it "stick" di sapone da barba
ne scheerstaaf
pl mydło do golenia w sztyfcie

521 shrink resistance
fr irrétrécissabilité
de Schrumpffestigkeit
el resistencia al encogimiento
it irrestringibilità
ne krimpbestendigheid
pl niekurczliwość, odporność na kurczenie
się

522 shrinkage
fr rétrécissement
de Schrumpfen, Krumpfen
el contracción, encogimiento
it restringimento

ne krimpen
pl skurcz, kurczenie się

523 sizing assistant
Product which, when added to sizing compounds, makes the warp yarns more flexible and slippery for the subsequent operation of weaving.
Note: These products may, for example, be sulphated or emulsified waxes and greases, possibly with the addition of wetting agents.

fr adjuvant d'encollage
Produit qui, ajouté aux compositions d'encollage, améliore la souplesse et le glissant des fils de chaîne pour l'opération ultérieure de tissage.
Nota: Il s'agit, par exemple, de graisses et cires sulfatées ou émulsionnées, éventuellement additionnées d'agents mouillants.

de Schlichtehilfsmittel
Produkt, das Schlichtemitteln zugesetzt wird. Es verbessert die Weichheit und Glätte der Kettfäden für den nachfolgenden Webprozess.
Anmerkung: Es handelt sich hierbei z.B. um sulfierte oder emulgierte Fette und Wachse, evtl. unter Zusatz von Netzmitteln.

el auxiliar de encolado
Producto que, añadido a las composiciones de encolado, mejora la suavidad y lisura de los hilos de urdimbre, para la operación consecutiva de tisaje.
Observación: Se trata, por ejemplo, de grasas y ceras sulfatadas o emulsionadas, adicionadas a veces de agentes humectantes.

it —

ne sterkhulpmiddel
Een produkt dat toegevoegd aan de sterkpap, de behandelde kettinggarens soepeler en gladder maakt t.b.v. het hierna volgende weefproces.

pl wspomagacz klejenia
Produkt dodawany do klejonek nadający przędzy osnowowej większą elastyczność i śliskość w prowadzonej następnie operacji tkania.
Uwaga: Produktami takimi mogą być np. siarczanowane lub zemulgowane woski i tłuszcze, ewentualnie z dodatkiem środków zwilżających.

524 skin colour
fr teint
de Hautfarbe
el cutis
it colorito
ne huidkleur
pl barwa skóry

525 skin condition
fr condition de la peau
de Hautbefund
el condición de la piel
it condizione della pelle
ne toestand van de huid
pl stan skóry

526 skin damage, skin irritation
fr dommage de la peau, irritation dermique, affection dermatologique, dermite, dermatose, lésion de la peau
de Hautschäden, Hautreizung
el dañado de la piel, irritación de la piel, afección dermatológica, dermatosis, lesión de la piel
it danno della pelle, irritazione della pelle, affezione dermatologica, dermite, dermatosi, lesione della pelle
ne beschadiging van de huid, huidirritatie
pl uszkodzenie skóry, podrażnienie skóry

527 skin irritation
fr irritation de la peau
de Haut-Irritation
el irritación de la piel
it irritazione della pelle
ne huidirritatie
pl podrażnienie skóry

528 skin lightener
fr crème pour éclaircir
de Hautaufheller
el crema para aclarar la piel
it crema per schiarire e rendere licia la pelle
ne middel om de huid lichter te maken
pl środek do rozjaśniania skóry

529 skin lotion
fr lotion pour la peau
de flüssige Hautcreme
el loción para la piel
it lozione per la pelle
ne toiletwater voor de huid
pl płyn do pielęgnacji skóry

530 skin protecting agent, skin protective
fr agent protecteur pour la peau, émollient
de Hautschutzmittel
el agente protector de la piel, emoliente
it agente protettore della pelle, emolliente
ne huidbeschermend middel
pl środek ochraniający skórę

531 skin protecting soap
Soap which contains special additives which protect the skin against occupational dermatitis.
fr savon pour protéger la peau
Savon contenant des additifs spéciaux

pour protéger la peau contre les derma-
toses professionnelles.

de spezielle Hautschutzseife
Seife mit besonderen, z.T. auch substan-
tiven Hautschutzstoffen, zum Schutz
gegen Berufsdermatosen, usw.

el jabón para protección de la piel
Jabón con substancias especiales, en
parte sustantivas para la piel, para su
protección contra las dermatosis pro-
fesionales, etc.

it sapone protettivo della pelle
Saponi che contengono speciali additivi
che proteggono la pelle contro possibili
dermatiti.

ne huidbeschermende zeep
Zeep met speciale toevoegsels, die de
huid tegen beroepseczemen beschermt.

pl mydło ochronne
Mydło zawierające specjalne dodatki
zabezpieczające skórę przed zawodowym
zapaleniem skóry.

532 soap

An alkaline salt (inorganic or organic) formed from a fatty acid or mixture
of fatty acids containing at least 8 carbon atoms. This anionic surface active
agent exhibits the phenomenon of
reversible hydrolysis by the action of
water. Because of this fact, water
soluble soaps or true soaps exhibit
characteristic properties; their reaction
is usually alkaline.

Note: 1. In practice, part of the fatty acids may
be replaced by resin acids.
2. Generally, the term "metallic soap" is limited
to the soaps of fatty acids of non alkaline metals.
In practice, these salts are insoluble in water
and do not possess detergent properties.

fr savon
Sel alcalin (inorganique ou organique)
d'un acide gras ou d'un mélange d'acides
gras contenant au moins huit atomes de
carbone. Cet agent de surface anionique,
par l'action de l'eau, donne lieu au
phénomène de l'hydrolyse réversible.
Les savons solubles dans l'eau, ou
savons proprement dits, manifestent de
ce fait des propriétés caractéristiques: ils
sont à réaction généralement alcaline.

Nota: 1. Dans la pratique, une partie des acides
gras peut être remplacée par des acides résiniques.
2. Dans l'usage courant, l'appellation "savon
métallique" est réservée aux sels d'acides gras
des métaux non-alcalins. Pratiquement, ces sels
sont insolubles dans l'eau et ne possèdent pas
de propriétés détergentes.

de Seife
Alkalisalz (anorganisches oder organisches) einer Fettsäure oder einer Mischung von Fettsäuren mit wenigstens 8
Kohlenstoffatomen. Diese anionische
grenzflächenaktive Verbindung zeigt mit
Wasser die Erscheinung der reversiblen
Hydrolyse. Die wasserlöslichen Seifen im engeren Sinn erhalten dadurch
ihre charakteristischen Eigenschaften:
Sie reagieren im allgemeinen alkalisch.

Anmerkung: 1. Ein Teil der Fettsäuren kann in
der Praxis durch Harzsäuren ersetzt werden.
2. Die Bezeichnung "Metallseife" ist im allgemeinen Sprachgebrauch den betr. Salzen der
Nichtalkalimetalle reserviert. Diese Salze sind
praktisch wasserunlöslich und zeigen keine wasch-
aktive Eigenschaften.

el jabón
Sal alcalina de un ácido graso o de una
mezcla de ácidos grasos, que contengan
por lo menos ocho átomos de carbono.
El jabón es un detergente de superficie
aniónico, que por la acción del agua da
lugar al fenómeno de la hidrólisis reversible. A causa de ella, los jabones solubles en agua, o jabones propiamente
dichos, poseen ciertas propiedades características, siendo su reacción generalmente alcalina.

Observaciones: 1. En la práctica, una parte de
los ácidos grasos puede ser reemplazada por
ácidos resínicos.
2. En el uso corriente, la denominación "jabón
metálico" se reserva a las sales de ácidos grasos
de metales no alcalinos. Prácticamente, estas
sales son insolubles en agua y no poseen propiedades detergentes.

it sapone
Sale alcalino di una miscela di acidi
grassi contenenti almeno 8 atomi di
carbonio. Questo tensioattivo anionico
dà luogo, per azione dell'acqua, al
fenomeno dell'idrolisi reversibile. Come
conseguenza i saponi solubili in acqua
o saponi propriamente detti presentano
proprietà caratteristiche: la loro reazione
è normalmente alcalina.

Nota: In pratica, parte degli acidi grassi può essere
sostituita da acidi resinici.

ne zeep
Een basisch zout gevormd uit een
mengsel van vetzuren met ten minste 8
koolstofatomen (in de keten).
Deze anionaktieve oppervlakactieve stof
veroorzaakt het verschijnsel van omkeerbare hydrolyse door inwerking van
water. Als gevolg hiervan hebben in
water oplosbare zepen of echte zepen

karakteristieke eigenschappen; zij reageren gewoonlijk alkalisch.

Opmerking: 1. In de praktijk kan een deel der vetzuren vervangen worden door harszuren. *2.* De term "metaalzepen" wordt gebruikt voor de non-alkalizouten van vetzuren. Deze zouten zijn nagenoeg onoplosbaar in water en bezitten geen waswerking.

pl mydło
Sól zasady nieorganicznej lub organicznej otrzymana z kwasu tłuszczowego lub mieszaniny kwasów tłuszczowych zawierających co najmniej 8 atomów węgla. Ten anionowy związek powierzchniowo czynny wykazuje pod działaniem wody zjawisko odwracalnej hydrolizy. W związku z tym rozpuszczalne w wodzie mydła, tj. prawdziwe mydła, wykazują pewne charakterystyczne własności. Ich odczyn na ogół jest alkaliczny.

Uwagi: 1. W praktyce przemysłowej część kwasów tłuszczowych może być zastąpiona przez kwasy żywiczne. *2.* Określenie "mydła metaliczne" w zasadzie ograniczone jest do soli kwasów tłuszczowych i metali niealkalicznych. Sole te praktycznie są nierozpuszczalne w wodzie i nie wykazują własności piorących.

533 soap flakes
fr copeaux de savon, savons en paillettes
de Seifenflocken
el jabón en copos, jabón en escamas, jabón en virutas
it trucioli di sapone, saponi in scaglie
ne zeepvlokken
pl płatki mydlane

534 soap mill
fr boudineuse
de Piliermaschine
el cilindros para jabón, molino para jabón
it laminatrice a cilindri, raffinatrice a cilindri
ne pileermachine
pl ucieraczka walcowa

535 soap pan, soap boiler, soap kettle (USA)
fr cuve, chaudron
de Siedekessel, Siedepfanne
el caldera
it caldaia
ne zeeppan, zeepketel
pl kocioł warzelny

536 sodium bromate
fr bromate de sodium
de Natriumbromat
el bromato sódico
it bromato di sodio
ne natriumbromaat
pl bromian sodowy

537 soft soap
Semi-boiled soap obtained from liquid fatty oils and acids and potassium hydroxide.
fr savon mou
Savon mi-cuit obtenu avec des huiles ou acides gras liquides et de l'hydroxyde de potassium, contenant de 35 à 45% d'acides gras.
de Schmierseife
Leimseife, die aus Ölen oder flüssigen Fettsäuren durch Verseifung mit Kaliumhydroxyd hergestellt wird.
el jabón blando
Jabón de empaste en caliente, obtenido con aceites o ácidos grasos líquidos y con hidróxido potásico.
it sapone molle
Sapone d'impasto (sapone semicotto) ottenuto da acidi grassi e grassi di consistenza liquida (olii), in reazione con idrato di potassio.
ne zachte zeep
Lijmzeep, die uit oliën of vloeibare vetzuren met kaliumhydroxide bereid wordt.
pl mydło maziste
Mydło otrzymane z olejów i ciekłych kwasów tłuszczowych oraz wodorotlenku potasowego.

538 soft water
fr eau douce
de weiches Wasser
el agua dulce, agua blanda, agua poco dura
it acqua dolce
ne zacht water
pl woda miękka

539 soften
fr adoucir
de enthärten
el ablandar
it addolcire
ne ontharden
pl zmiękczać (wodę)

540 softening
fr adoucissement
de Enthärtung
el ablandamiento
it addolcimento
ne ontharding
pl zmiękczanie (wody)

541 softening agent
Product used to make the processed textile more flexible, and consequently to obtain a given feel. It is also used as an additive in sizing and finishing baths, etc.

Note: These products are generally surface active agents or preparations based on greases and oils with suitable emulsifying agents. Brightening agents may also be used.

fr agent d'adoucissage
Produit servant à augmenter la souplesse du textile traité, et par suite à obtenir un toucher déterminé. Il est utilisé également comme additif dans les bains d'encollage, d'apprêts, etc.

Nota: Il s'agit en général d'agents de surface ou de préparations à base de graisses et d'huiles avec des agents émulsionants appropriés. Des agents d'avivage peuvent être également utilisés.

de Weichmachungsmittel
Produkt, das zur Erhöhung der Geschmeidigkeit und damit zur Erzielung eines bestimmten Griffcharakters der zu behandelnden Textilien dient. Es findet auch als Zusatz zu Schlichtflotten, Appreturflotten usw. Verwendung.

Anmerkung: Es handelt sich in der Regel um grenzflächenaktive Stoffe oder Zubereitungen von Fetten und Ölen mit geeigneten Emulgiermitteln. Avivagemittel können auch verwendet werden.

el agente suavizante
Producto que sirve para dar a la materia textil un tacto suave y una variación de tacto determinado (lisura, cuerpo, crujido). También se utiliza como aditivo en los baños de encolado, de apresto, etc.

Observación: Se trata en general, de los mismos agentes de superficie o de composiciones a base de grasas y de aceites con agentes emulsionantes apropiados. También pueden utilizarse compuestos de amonio cuaternario.

it —

ne zacht- resp. weekmaker
Een produkt dat gebruikt wordt om garen of doek flexibeler te maken waardoor de gewenste greep wordt verkregen.

pl środek zmiękczający
Środek stosowany do nadania wyrobom włókienniczym większej elastyczności, a więc w rezultacie właściwego chwytu. Stosowany jest on również jako dodatek do kąpieli klejarskich, kąpieli wykończalniczych itd.

Uwaga: Produktami tego typu są na ogół związki i powierzchniowo czynne lub preparaty oparte na tłuszczach i olejach z dodatkiem odpowiednich emulgatorów. Stosowane mogą być również środki ożywiające.

542 softening effect
fr effet adoucissant
de enthärtende Wirkung
el efecto desalado, efecto del permutado
it effetto di addolcimento
ne ontharde werking
pl działanie zmiękczające, zmiękczanie

543 soil
The undesirable deposit on the surface and/or within the substrate, which changes some characteristics of appearance or feel of clean surfaces.

fr salissure (souillure)
Apport indésirable, en surface et/ou à l'intérieur du substrat, altérant certains caractères d'aspect ou de toucher des surfaces propres.

de Schmutz
Unerwünschte Auflage auf der Oberfläche und/oder im Innern des Substrats, die gewisse Eigenschaften wie Aussehen oder Griff der sauberen Oberfläche verändert.

el suciedad
Depósito indeseado, en superficie y/o en el interior del substrato, que altera ciertas características de aspecto o de tacto de las superficies limpias.

it sporco
Sostanze indesiderabili in superficie e/o all'interno di un substrato, che ne alterano l'aspetto ed il tatto.

ne vuil
Het ongewenste neerslag op een oppervlak of op een substraat, dat sommige uiterlijke kenmerken en het aanvoelen daarvan beïnvloedt.

pl brud
Niepożądany osad na powierzchni substratu i/lub wewnątrz niego, zmieniający pewne własności, takie jak wygląd i dotyk czystej powierzchni.

544 soil adherence
fr persistance de la salissure, affinité de la souillure pour la fibre, adhérence de la salissure
de Schmutzhaftfestigkeit
el persistencia de la suciedad, adherencia de la suciedad, afinidad de la suciedad para la fibra
it persistenza dello sporco, affinità sporco per la fibra, aderenza dello sporco
ne affiniteit van het vuil tot de vezel
pl przyczepność brudu, przyleganie brudu

545 soil redeposition
fr redéposition des salissures
de Wiederaufziehen von Schmutz
el redeposición de la suciedad
it rideposizione dello sporco
ne weer neerslaan van het vuil
pl wtórne osadzanie się brudu, redepozycja brudu

546 soil removing capacity
fr pouvoir d'enlèvement de la salissure, pouvoir d'élimination de la salissure, pouvoir de disperser les salissures
de Schmutzlösevermögen, Schmutzentfernungsvermögen
el poder de eliminación de la suciedad, poder dispersante de la suciedad
it potere di rimuovere lo sporco, potere di eliminazione dello sporco
ne vuilverwijderend vermogen
pl zdolność usuwania brudu

547 soil residue, residual soil, remaining soil
fr crasse restante, souillure restante
de Restschmutz
el suciedad restante, residuo de la suciedad, suciedad residual
it residuo di sporco
ne resterend vuil
pl pozostałosć brudu, zabrudzenie szczątkowe

548 soil suspending power
fr pouvoir suspensif
de Schmutztragevermögen
el poder suspensor de la suciedad, poder de suspensión de la suciedad
it potere sospensivo
ne vuildragend vermogen
pl zdolność odtransportowywania brudu

549 soiling composition
fr composition de la souillure
de Schmutzzusammensetzung
el composición de la suciedad
it composizione dello sporco
ne samenstelling van het vuil
pl skład zabrudzenia

550 soiling operation, soiling procedure
fr salissement
de Anschmutzen
el ensuciar
it operazione di sporcare
ne bevuiling
pl brudzenie

551 solid lime, soap scum, scum
fr crasses insolubles, sels calcaires insolubles, écume calcaire
de unlösliche Kalkseife
el sales calcáreas insolubles
it residui insolubili, sali calcarei insolubili
ne onoplosbare kalkzeep
pl zwarzaki mydła wapniowego

552 solubilisation†
fr solubilisation
de Solubilisierung
el solubilización
it solubilizzazione
ne solubilisatie
pl solubilizacja
   † See appendix

553 solubiliser, hydrotrope
fr agent de solubilisation, tiers solvant
de Lösungsvermittler
el agente solubilizante, fracción disolvente
it agente di solubilizzazione
ne solubiliserend agens
pl hydrotrop
   Substancja zwiększająca rozpuszczalność związków organicznych w wodzie.

554 solubilizing and/or dispersing agent for dyestuffs
Product promoting the solubilization and/or the aqueous dispersion of dyes and consequently improving their dyeing properties (efficiency, penetration).
Note: These products are surface active agents, for example: esters and amides of sulphated fatty acids, fatty acid condensates, alkylarylsulphonates, oxalkylated products, derivatives of aliphatic amines.

fr agent de dissolution et/ou de dispersion des colorants
   Produit favorisant la dissolution et/ou la dispersion aqueuse des colorants et améliorant par suite leurs propriétés tinctoriales (rendement, pénétration, etc.).
   Nota: Il s'agit d'agents de surface, par exemple: esters et amides d'acides gras sulfatés, condensats d'acides gras, alkylarylsulfonates, produits d'oxyalkylation, dérivés d'amines aliphatiques.

de Farbstofflöse- und/oder -dispergiermittel
   Produkt, das das Auflösen und/oder Dispergierung der Farbstoffe im Wasser unterstützt und damit die Voraussetzung für einen einwandfreien Färbeverlauf schafft (Ausbeute, Durchfärbevermögen usw.).
   Anmerkung: Es handelt sich um grenzflächenaktive Stoffe, beispielsweise um sulfierte Fettsäureester und -amide, Fettsäurekondensationsprodukte,

Alkylarylsulfonate, Oxalkylierungsprodukte sowie Derivate aliphatischer Amine.

el agente de disolución y/o de dispersión de colorantes
Producto que favorece la disolución y/o la dispersión en agua de los colorantes, mejorando como consecuencia sus propiedades tintóreas (rendimiento, poder de penetración, etc.).
Observación: Se trata de agentes de superficie, con o sin disolventes, por ejemplo: ésteres y amidos de ácidos grasos sulfatados, productos de condensación de ácidos grasos, alquilarilsulfonatos, ésteres y éteres de poliglicoles, así como derivados de aminas alifáticas.

it —

ne dispergeermiddel voor verfstoffen
Een produkt dat de dispersie van kleurstoffen bevordert en/of een bepaalde dispersiteitsgraad in stand houdt door aggregatie of bezinking tegen te gaan.

pl środek solubilizujący i/lub dyspergujący barwniki
Środek poprawiający rozpuszczanie i/lub dyspergowanie barwników w wodzie, w wyniku czego poprawiają się ich własności barwiące (wydajność, penetracja).
Uwaga: Są to związki powierzchniowo czynne, takie jak np. estry i amidy siarczanowanych kwasów tłuszczowych, produkty kondensacji kwasów tłuszczowych, alkiloarylosulfoniany, produkty oksyalkilenowania oraz pochodne amin alifatycznych.

555 solubilizing and/or dispersing agent for pigments
Product covered by the definition corresponding to the Solubilizing and/or dispersing agent for dyestuffs.

fr agent de dissolution et/ou de dispersion des pigments
Produit auquel s'applique la définition correspondant à Agent de dissolution et/ou de dispersion des colorants pour la teinture.

de Farbstofflöse- und/oder -dispergiermittel
Produkt, für das die Definition zu Farbstofflöse- und/oder -dispergiermittel für die Färberei zutrifft.

el agente de disolución y/o de dispersión de pigmentos
Producto al cual se aplica la definición correspondiente a agente de disolución y/o de dispersión de colorantes de tintura.

it —

ne solubilisator en/of dispergator voor pigmenten
Produkt waarvoor de definitie van het

solubiliseer- en/of dispergeermiddel voor verfstoffen van toepassing is.

pl środek solubilizujący i/lub dyspergujący pigmenty
Produkt objęty definicją środka solubilizującego i/lub dyspergującego barwniki.

556 solubilizing power
The extent of the ability of a dissolved surface active agent to confer on certain bodies, of low solubility in the pure solvent, an apparent solubility by micelle formation.

fr pouvoir de solubilisation
Degré d'aptitude d'un agent de surface en solution à donner à certains corps peu solubles dans le solvant pur une solubilité apparente, par association micellaire.

de Solubilisiervermögen
Grad der Fähigkeit einer gelösten grenzflächenaktiven Verbindung, gewissen, im reinen Lösungsmittel schwer löslichen Verbindungen durch micellare Assoziation eine scheinbare Löslichkeit zu vermitteln.

el poder de solubilización
Capacidad de un agente de superficie en disolución para dar, por asociación micelar, una solubilidad aparente a ciertas substancias, poco solubles en el disolvente puro.

it potere solubilizzante
Grado dell'attitudine di un tensioattivo in soluzione a conferire a determinate sostanze poco solubili nel solvente puro una solubilità apparente per associazione micellare.

ne solubiliseervermogen
Het vermogen van een opgeloste oppervlakaktieve stof bepaalde verbindingen met geringe oplosbaarheid een schijnbare oplosbaarheid te verlenen door de vorming van micellen.

pl zdolność solubilizacji
Zakres zdolności rozpuszczonego związku powierzchniowo czynnego do nadawania substancjom trudno rozpuszczalnym w czystych rozpuszczalnikach pozornej rozpuszczalności dzięki tworzeniu miceli.

557 special soaps
Soap produced by saponification with alkaline or alkaline earth hydroxides, ammonia, organic amines or by double decomposition of alkaline soaps with salts of heavy metals. They are used in

the field of cosmetics or for technical uses, as the metallic soaps.

fr savons spéciaux
Savon obtenu par saponification avec des hydroxydes alcalins ou alcalino-terreux, ammoniaque ou amines organiques ou par double décomposition de savons alcalins avec des sels des métaux lourds, qui s'emploient dans le domaine de la cosmétique ou pour des usages techniques, comme les savons métalliques.

de Spezialseifen
Seifen, die durch Verseifung mit Alkalien, Erdalkalien, Ammoniak oder organischen Aminen oder durch Umsetzung von Alkali-Seifen mit Schwermetallsalzen hergestellt werden und auf dem Gebiet der Körperpflege bzw. Hygiene oder auf technischem Gebiet Verwendung finden, wie z.B. Metallseifen.

el jabones especiales
Jabones obtenidos por saponificación con hidróxidos alcalinos, alcalino-térreos, amoníaco, aminas o por doble descomposición de jabones alcalinos con sales de metales pesados y que se emplean en el campo de la cosmética para el cuidado del cuerpo y de la higiene o en el campo de la técnica, como los jabones abrasivos, jabones con disolventes, quita-manchas y los jabones metálicos.

it sapone speciale
Saponi prodotti a mezzo saponificazione con idrossidi alcalini o alcalino-terrosi, ammoniaca, ammine organiche o dalla reazione di doppio scambio fra saponi alcalini e sali di metalli pesanti. Sono usati in cosmetica o per usi tecnici, come saponi metallici.

ne speciale zepen
Zepen, die door verzeping met alkaliën, aardalkaliën, ammoniak of organische aminen of door omzetting van alkali-zepen met zouten van zware metalen vervaardigd worden en op kosmetisch of technisch gebied toepassing vinden, zoals b.v. metaalzepen.

pl mydła specjalne
Mydła otrzymane bądź przez zmydlenie za pomocą alkaliów lub wodorotlenków metali ziem alkalicznych, amoniaku lub amin organicznych, bądź w reakcji podwójnej wymiany między mydłami alkalicznymi a solami metali ciężkich. Znaj-

dują one zastosowanie w kosmetyce, jak również w technice jako mydła metaliczne.

558 spinning bath additive
Product used, among other things, for clarifying the spinning bath and preventing the nozzles and conduits from becoming clogged.
Note: These products are generally surface active agents or preparations comprising them, for example: sulphated oils, alkylsulphonates, fatty acid condensates, oxyethyl alkylamines, onium derivatives.

fr additif pour bain de filage
Produit servant, entre autres, à clarifier le bain de filage et à éviter le bouchage des filières et des conduits.
Nota: Il s'agit en général d'agents de surface ou de préparations en comportant, par exemple: huiles sulfatées, alkylsulfonates, condensats d'acides gras, alkylamines oxyéthylées, dérivés d'ammonium quaternaire.

de Spinnbadzusatzmittel
Produkt, das unter anderem dazu dient, das Spinnbad zu klären und das Zuwachsen der Spinndüsen und Spinnbadleitungen zu verhindern.
Anmerkung: Es handelt sich im allgemeinen um grenzflächenaktive Stoffe oder Zubereitungen hieraus, z.B. um sulfierte Öle, Alkylsulfonate, Fettsäurekondensationsprodukte, Äthoxylierungsprodukte von Alkylaminen sowie Oniumverbindungen.

el aditivo para el baño de hilatura
Producto que, entre otras aplicaciones, sirve para clarificar el baño de hilatura y para evitar el taponamiento de las hileras.
Observación: Se trata en general, de agentes de superficie, o de preparaciones que los contienen. Se utilizan aceites sulfatados, alquilsulfonatos, productos de condensación de ácidos grasos, alquilaminas oxietiladas, así como derivados onio.

it —

ne spinbadadditief
Dit produkt wordt o.a. gebruikt voor het klaren van het spinbad en tegen het verstopt raken van sproeiers.
Opmerking: Deze produkten zijn gewoonlijk oppervlakaktieve stoffen of preparaten waarin zij verwerkt zijn, b.v. gesulfateerde oliën, alkylsulfonaten, vetzuurcondensaten, ethoxyalkylaminen, quaternaire ammonium verbindingen.

pl dodatek do kąpieli przędzalniczych
Produkt stosowany między innymi do klarowania kąpieli przędzalniczych oraz zabezpieczania dysz przędzalniczych przed zatykaniem się.
Uwaga: Produktami takimi są na ogół związki powierzchniowo czynne, takie jak siarczanowane oleje, alkilosulfoniany, produkty kondensacji kwasów tłuszczowych, oksyetylenowane alkiloaminy, związki oniowe oraz preparaty zawierające te związki.

559 spinning oil[1]

Product which, applied to the fibres in the course of their preparation for spinning, makes them more slippery and flexible and possibly gives them other surface qualities (e.g., cohesion), with a view to the operations of combing and spinning. Depending on the purpose for which they are used, spinning oils may also possess the properties of wetting agents and fulling asistants and other secondary properties, e.g. that of promoting the sweating of hard or stem fibres[2].

Notes: 1. A spinning oil is a preparation agent more specifically suitable for spinning.
2. These products are mainly oil or grease based preparations possibly associated with emulsifying agents or with special surface active agents.

fr   agent d'ensimage[3]

Produit qui, appliqué sur les fibres au cours des opérations de préparation à la filature, les rend plus glissantes, plus souples, et peut leur conférer d'autres qualités de surface (cohésion, par exemple), en vue des opérations d'étirage et de filature. Selon son but d'utilisation, l'agent d'ensimage peut également posséder des propriétés de mouillant, d'adjuvant de foulage, et d'autres propriétés secondaires, par exemple, favoriser l'échauffe des fibres dures ou libériennes[4].

Nota: 1. L'agent d'ensimage est un agent de préparation plus spécialement adapté à la filature.
2. Il s'agit principalement de préparations à base d'huiles ou de graisses éventuellement associées à des agents émulsionnants ou à des agents de surface spéciaux (alkylphosphates, par exemple).

de   Schmälz- und Batschmittel[5]

Produkt, das vor dem Verspinnen auf das Fasergut aufgebracht wird und diesem einerseits Glätte und Geschmeidigkeit, sowie andererseits eine für das Verstrecken und Verspinnen günstige Oberflächenbeschaffenheit (z.B. besseren Zusammenhalt der Fasern) verleiht. Darüber hinaus können Schmälzmittel, je nach ihrem Verwendungszweck, auch netzende oder das Walken fördernde Eigenschaften besitzen und Sekundäreffekte, z.B. beim Röten von Hart- und Bastfasern begünstigen[6].

Anmerkung: 1. Schmälzmittel sind Präparationsmittel, die speziell dem Spinnprozess angepasst sind.
2. Es handelt sich überwiegend um Zubereitungen von zum Teil auch abgewandelten oder um spezielle grenzflächenaktive Stoffe (wie z.B. Phosphorsäureester).

el   agente de ensimaje (aceite de ensimaje, aceite de deshilachado)

Producto que aplicado a las fibras durante las operaciones de preparación a la hilatura, las da lisura y ductilidad, pudiendo además conferirles, a causa de sus propiedades adherentes, una mejor coherencia durante las operaciones de estirado y de hilatura. Según su finalidad el agente de ensimaje puede poseer igualmente propiedades humectantes, o que favorezcan el batanado, así como efectos secundarios, por ejemplo, en el macerado de fibras duras y liberianas (llamados en Alemania "Batschmittel" y en Gran Bretaña "batching oil").

Observaciones: 1. El agente de ensimaje es un agente de preparación especialmente adaptado a la hilatura.
2. Se trata principalmente de preparaciones a base de aceites o de grasas, asociadas a veces con agentes emulsionantes, o de agentes de superficie especiales.

it   —

ne   spinavivage

Produkten welke toegepast worden om de produktie van synthetische vezels mogelijk te maken (spinnen), of de verwerking van vezels tot garen (verspinnen) mogelijk te maken.

Opmerking: 1. Spinolie
Een produkt dat voor het verspinnen, op de vezels wordt gebracht met het doel de verspinbaarheid te verbeteren of mogelijk te maken.
2. Smoutmiddel
Naam voor spinolie gangbaar in de wolsektor.

pl   natłustka

Produkt nanoszony na włókna podczas przygotowywania ich do przędzenia, służący do nadania włóknu śliskości i elastyczności. Produkt ten może służyć również do nadania powierzchni własności pożądanych przy operacjach czesania i przędzenia, np. kohezji. W zależności od przeznaczenia natłustki mogą posiadać również inne własności dodatkowe takie jak wspomaganie roszenia włókien liściowych i łykowych.

Uwagi: 1. Natłustka jest preparatem specjalnie przeznaczonym do przędzenia.
2. Są to głównie preparaty oparte na olejach lub tłuszczach, ewentualnie spokrewnione ze środkami emulgującymi lub specjalnego typu związkami powierzchniowo czynnymi.

---

[1] Spinning oils are also called batching oil, tearing oil, etc...
[2] Products of this type used for the sweating and spinning of hard or stem fibres are called, in Germany and Great-Britain respectively, "Batschmittel" and "batching oil".
[3] Les agents d'ensimage sont également appelés huile d'ensimage, huile d'effilochage, etc.
[4] Les produits de ce genre, utilisés pour l'échauffe

et la filature des fibres dures ou libériennes, sont appelés, en Allemagne et en Grande-Bretagne, respectivement: "Batschmittel" et "batching oil".
5) Schmälzmittel werden auch Schmälzöle, Reissöle, usw. genannt.
6) Produkte dieser Art, die beim Rösten und Verspinnen von Hart- bzw. Bastfasern angewandt werden, heissen in Deutschland und in Grossbritannien "Batschmittel" bzw. "batching oil".

560 spinning solution additive
Product added during the preparation of the spinning solution, for the purpose of improving the suitability of the solution for spinning and possibly to alter the quality of the filaments.

Note: These products are generally surface active agents or preparations comprising them, for example: sulphated oils, alkylsulphates, fatty acid condensates, oxyethyl alkylamines, onium derivatives.

fr additif pour solution de filage
Produit ajouté au cours de la préparation de la solution de filage, destiné à améliorer l'aptitude au filage de la solution et à modifier éventuellement la qualité des filaments.

Nota: Il s'agit en général d'agents de surface ou de préparations en comportant, par exemple: huiles sulfatées, alkylsulfates, condensats d'acides gras, alkylamines éthoxylées, dérivés onium.

de Zusatzmittel zu Spinnlösungen
Produkt, das bei der Herstellung der Spinnlösung zur Verbesserung der Verspinnbarkeit zugesetzt wird und gegebenenfalls die Eigenschaften des Spinnerzeugnisses modifiziert.

Anmerkung: Es handelt sich in der Regel um grenzflächenaktive Stoffe oder Zubereitungen hieraus, z.B. sulfierte Öle, Alkylsulfonate, Fettsäurekondensationsprodukte, Äthoxylierungsprodukte von Alkylaminen sowie Oniumverbindungen.

el aditivo para la disolución de hilatura
Producto añadido en la preparación de la disolución de hilatura, para mejorar la aptitud a la hilatura de la disolución y en algunos casos para modificar las propiedades de los filamentos.

Observación: Se trata en general, de agentes de superficie, o de preparaciones que los contienen. Se utilizan aceites sulfatados, alquilsulfónicos, productos de condensación de ácidos grasos, alquilaminas oxietiladas, así como derivados onio.

it —
ne spinoplossingsadditief
Produkt dat wordt toegevoegd tijdens de bereiding van de spinoplossing met als doel de geschiktheid van de oplossing voor het spinnen te verbeteren en eventueel de hoedanigheid van de filamenten te veranderen.

Opmerking: Deze produkten zijn gewoonlijk oppervlakaktieve stoffen of preparaten waarin zij verwerkt zijn, b.v. gesulfateerde oliën, alkylsulfonaten, vetzuurcondensaten, ethoxylalkylaminen, quaternaire ammoniumverbindingen.

pl dodatek do roztworu przędnego
Produkt dodawany podczas przygotowywania roztworu przędnego dla poprawienia jego przydatności do przędzenia oraz ewentualnie dla zmodyfikowania włókna.

Uwaga: Produktami takimi są na ogół związki powierzchniowo czynne takie jak siarczanowane oleje, alkilosiarczany, produkty kondensacji kwasów tłuszczowych, oksyetylenowane alkiloamidy, pochodne oniowe oraz preparaty zawierające te związki.

561 sponge
fr éponge
de Schwamm
el esponja
it spugna
ne spons
pl gąbka

562 spotting agent
Product intended to remove local stains on textile articles. A distinction is drawn between "dry" and "wet" spotting agents, depending on whether they act in a solvent or an aqueous medium.

Note: These products are mainly preparations based on solvents and surface active agents with emulsifying and detergent properties, such as: amine soaps, alkylsulphates, alcane sulphonates, fatty acid condensates, alkylarylsulphonates, oxalkylated products.

fr agent détachant
Produit destiné à éliminer les salissures locales sur des articles textiles. On distingue les agents détachants "à sec" et les agents détachants "au mouillé" selon qu'ils agissent en milieu solvant ou aqueux.

Nota: Il s'agit essentiellement de préparations à base de solvants et d'agents de surface ayant des propriétés émulsionnantes et détergentes, tels que: savons d'amines, alkylsulfates, alkylsulfonates, condensats d'acides gras, alkylarylsulfonates, produits oxyalkylés.

de Detachiermittel
Produkt, das zur Entfernung örtlich begrenzter Verschmutzungen von Textilwaren bestimmt ist. Man unterscheidet je nach Einsatz in Lösungsmitteln oder im wässrigen Medium Trocken- und Nassdetachiermittel.

Anmerkung: Es handelt sich im wesentlichen um Kombinationen aus Lösungsmitteln und grenzflächenaktiven Stoffen mit emulgierenden und waschenden Eigenschaften, wie Aminseifen, Alkylsulfate, Alkansulfonate, Fettsäurekondensationsprodukte, Alkylarylsulfonate oder Oxalkylierungsprodukte.

el agente quitamanchas
Producto destinado a eliminar las manchas que aparecen localmente en los artículos textiles. Existen agentes quitamanchas "en seco" y "en húmedo", según que actúen con disolventes o en medio acuoso.

Observación: Se trata principalmente de composiciones de disolventes y agentes de superficie, con propiedades emulsionantes y detergentes, como jabones de aminas, alquilsulfatos, alquilsulfonatos, productos de condensación de ácidos grasos, alquilarilsulfonatos, ésteres y éteres de poliglicoles.

it —

ne detacheermiddel
Dit produkt heeft tot doel het verwijderen van plaatselijke vlekken op weefsels. Er wordt onderscheid gemaakt tussen "droge" en "natte" detacheermiddelen, al naar gelang zij in een oplosmiddel of in een waterig milieu werkzaam zijn.

pl środek odplamiający
Produkt do usuwania miejscowych zaplamień z wyrobów włókienniczych. Rozróżnia się "suche" i "mokre" środki odplamiające, zależnie od tego czy działają one w rozpuszczalniku czy też w środowisku wodnym.

Uwaga: Produkty tego typu są na ogół preparatami opartymi na rozpuszczalnikach i związkach powierzchniowo czynnych o własnościach emulgujących i piorących takich jak mydła aminowe, alkilosiarczany, alkilosulfoniany, produkty kondensacji kwasów tłuszczowych, alkiloarylosulfoniany i produkty oksyalkilenowane.

563 spray cooling
fr atomisation en cristallisation, atomisation à froid, séchage par l'air froid
de Kaltsprühung
el atomización por enfriamiento, secado por aire frío
it atomizzazione a freddo, essiccamento ad aria fredda
ne koud verstuiven
pl chłodzenie rozpyłowe

564 spray-dried household detergent product
fr détergent ménager atomisé
de sprühgetrocknetes Haushaltwaschmittel
el detergente doméstico atomizado
it detergente atomizzato ad uso domestico
ne verstoven huishoudwasmiddel
pl środek piorący suszony rozpyłowo przeznaczony do użytku w gospodarstwie domowym.

565 spray-drying
fr séchage par atomisation

de Sprühtrocknen, Sprühtrocknung, Zerstäubungstrocknung
el secado por atomización
it essiccazione a spruzzo o per atomizzazione
ne verstuivingsdroging
pl suszenie rozpyłowe

566 spray drying tower
fr tour d'atomisation, tour de pulvérisation, installation de séchage par atomisation, sécheur atomiseur
de Sprühturm, Trockenturm, Zerstäubungstrockner, Sprühanlage
el torre de atomización, torre de pulverización, instalación de secado por pulverización
it torre di atomizzazione
ne verstuivingsinstallatie, sproeitoren
pl wieża rozpyłowa susząca

567 spray drying with hot air, hot spraying
fr séchage par l'air chaud, atomisation à chaud, atomisation en séchage
de Heisssprühverfahren
el secado por aire caliente, atomización por calor, atomización en caliente
it essiccamento ad aria calda, atomizzazione a caldo
ne heet verstuiven
pl suszenie rozpyłowe gorącym powietrzem

568 spray nozzle
fr buse, gicleur
de Zerstäubungsdüse
el pulverizador
it ugello
ne drukverstuiver
pl dysza rozpyłowa

569 spreading ability
The property of a liquid, particularly of a solution of surface active agents, enabling a drop of this liquid to cover spontaneously another liquid or solid surface.

fr étalement
Propriété d'un liquide, et en particulier d'une solution d'agents de surface, permettant à une goutte de ce liquide de recouvrir spontanément une surface d'un autre liquide ou solide.

de Spreitung
Eigenschaft einer Flüssigkeit und im besonderen einer Lösung grenzflächenaktiver Verbindungen, die es einem Tropfen dieser Flüssigkeit ermöglicht,

eine Oberfläche einer anderen Flüssigkeit oder eines festen Körpers spontan zu bedecken.

el esparcimiento
Propiedad de un líquido y en particular de una disolución de agentes de superficie, por la que una gota de este líquido puede recubrir espontáneamente la superficie de otro líquido o de un sólido.

it spandimento
Proprietà di un liquido e in particolare di una soluzione di tensioattivi, per cui una sua goccia si spande spontaneamente sulla superficie di un altro liquido o di un solido.

ne spreidingsvermogen
De eigenschap van een vloeistof, in het bijzonder van een oplossing van oppervlakaktieve stoffen, die maakt dat een druppel van deze vloeistof een oppervlak van een vaste stof of een andere vloeistof spontaan bedekt.

pl zdolność rozlewania się, zdolność rozprzestrzeniania się
Właściwość cieczy, w szczególności roztworów związków powierzchniowo czynnych, umożliwiająca kropli tej cieczy pokrycie w sposób samorzutny powierzchni innej cieczy lub ciała stałego.

570 spreading tension, spreading coefficient
The tendency of a liquid to spread on a solid surface is expressed by the difference between the wetting tension and the surface tension of the liquid. This value is called spreading tension. It is expressed in newtons per metre (N/m)[1]. When the spreading tension is positive, the liquid spreads spontaneously on the solid surface.

fr tension d'étalement, coefficient d'étalement
La tendance d'un liquide à s'étaler sur une surface solide s'exprime par la différence entre la tension de mouillage et la tension superficielle du liquide. Cette grandeur s'appelle la tension d'étalement: elle s'exprime en newtons par mètre (N/m)[1]. Quand la tension d'étalement est positive, le liquide s'étale spontanément sur la surface solide.

de Spreitungsspannung (flüssig-fest), Spreitungskoeffizient
Das Bestreben einer Flüssigkeit, sich auf einer festen Oberfläche auszubreiten,

wird durch die Differenz zwischen der Benetzungsspannung und der Oberflächenspannung der Flüssigkeit ausgedrückt. Diese Grösse bezeichnet man als Spreitungsspannung. Sie wird in Newton pro Meter (N/m) ausgedrückt[1]. Ist der Spreitungskoeffizient positiv, so spreitet die Flüssigkeit spontan auf der Oberfläche.

el tensión de esparcimiento (líquido–sólido), coeficiente de esparcimiento
Tendencia de un líquido a esparcirse por una superficie sólida, representada por la diferencia entre la tensión de mojado y la tensión superficial del líquido. Cuando la tensión de esparcimiento es positiva, el líquido se esparce espontáneamente por la superficie sólida.

it tensione di spandimento (liquido–solido)
Tendenza di un liquido a spandersi su una superficie solida, espressa dalla differenza fra la tensione di bagnatura e la tensione superficiale del liquido. Questa grandezza si chiama tensione di spandimento e si esprime in dine per centimetro. Quando la tensione di spandimento è positiva il liquido si spande spontaneamente sulla superficie solida.

ne spreidingsspanning
De neiging van een vloeistof zich over een oppervlak van een vaste stof te verspreiden wordt uitgedrukt als het verschil tussen de spanning van het bevochtigen en de oppervlakspanning van de vloeistof. Deze waarde wordt de spreidingsspanning genoemd en wordt uitgedrukt in N/m[1]. Als de spreidingsspanning positief is, verspreidt de vloeistof zich spontaan over het vaste oppervlak.

pl współczynnik rozlewania się, współczynnik rozprzestrzenienia się
Dążność cieczy do rozprzestrzeniania się na powierzchni ciała stałego wyrażana jest różnicą pomiędzy jej napięciem zwilżania a jej napięciem powierzchniowym. Wielkość ta określana jest jako współczynnik rozlewania się i mierzona w niutonach na metr (N/m). Gdy współczynnik rozlewania się jest dodatni, ciecz rozlewa się na powierzchni ciała stałego w sposób samorzutny.
[1] $1\ N/m = 10^3\ dynes/cm$

571 stain removal
fr détachage, élimination des taches
de Fleckenentfernung
el quitar manchas, quitamanchas
it smacchiare, eliminazione delle macchie
ne vlekverwijdering, verwijderen van vlekken
pl usuwanie plam

572 stamping machine, stamper
fr presse-frappeuse, estampeuse
de Stempelmaschine
el troqueladora, estampadora
it stampatrice
ne stempelmachine
pl stemplarka

573 standard soil
fr souillure étalon, salissure standard, souillure normalisée
de standardisierter Schmutz
el suciedad estandarizada, suciedad normalizada
it sporco standard
ne standaardvuil
pl zabrudzenie standardowe

574 standard soiled fabric
fr tissu sali (normalisé standard)
de angeschmutzter standardisierter Teststoff
el tejido ensuciado estandarizado
it tessuto sporcato tipo standard
ne bevuilde standaardproefdoek
pl tkanina zabrudzona standardowo

575 static charge
fr charge statique
de statische Ladung
el carga estática
it carica statica
ne statische lading
pl ładunek elektryczności statycznej

576 stick
fr bâton
de Stift
el barrita
it stick
ne stift
pl sztyft, pałeczka

577 stick of colouring, block of colouring, (coloured) crayon for the hair
fr crayon colorant, pastel pour les cheveux
de Färbestift
el lápiz colorante
it matita per tingere i capelli

ne kleurstift
pl pastel koloryzujący, kredka koloryzująca (do włosów)

578 storage
fr stockage
de Lagerung
el almacenamiento, almacenaje
it stoccaggio
ne opslaan
pl magazynowanie, przechowywanie, składowanie

579 strengthening of the hair
fr consolidation du cheveu
de Haarverfestigung
el consolidación del cabello
it rafforzamento del capello
ne versterken van het haar
pl wzmacnianie włosów

580 stretching, elongation
fr allongement
de Streckung, Dehnung
el alargado, estirado, dilatación
it allungamento, elongazione
ne rek
pl rozciąganie, wydłużenie

581 stripping agent (partial or total)
A partial stripping agent is intended to lighten a dye which is too dark. It acts by eliminating part of the dye. Levelling agents are suitable for this operation. A total stripping agent, or dye removing agent, is used to eliminate the dye from a dyed fabric. Reducing agents are generally used for this, in conjunction with levelling agents.

fr agent de démontage (partiel ou total)
Un agent de démontage partiel est destiné à éclaircir une teinture trop foncée. Son action consiste à éliminer une partie du colorant. Les agents égalisants conviennent à cette opération. Un agent de démontage total, ou de décoloration, sert à éliminer le colorant d'un textile teint. Pour ce faire, on prend en général des agents réducteurs associés à des agents égalisants.

de Aufhellungs- und Abziehmittel
Ein Aufhellungsmittel hat den Zweck, zu dunkel ausgefallene Färbungen aufzuhellen. Es wirkt dadurch, dass es den Farbstoff teilweise abzieht. Hierfür eignen sich Egalisiermittel. Unter die Bezeichnung "Aufhellungsmittel" fallen nicht

die optischen Aufheller (Weisstöner).
Ein Abziehmittel dient dazu, gefärbte
Textilien zu entfärben. Für diesen Zweck
werden Reduktionsmittel in Kombination mit Egalisiermitteln verwendet.

el agente de desmontado (parcial o total)
Un agente de desmontado parcial se
destina a aclarar una tintura demasiado
oscura. Su acción consiste en eliminar
una parte del colorante. Los agentes
igualadores son adecuados para esta
operación. Un agente de desmontado
total, o de decoloración, sirve para
eliminar el colorante de una materia
textil teñida. Se emplean en general,
agentes reductores, asociados a agentes
igualadores.

it —

ne aftrekmiddel
Een produkt dat door chemische reactie
of door oplossende werking kleurstof
van textiel verwijdert. (Een partieel
aftrekmiddel dient om een verving, die
te donker is, lichter te maken.)

pl środek rozjaśniający lub środek odbarwiający (obciągający)
Środek rozjaśniający służy do rozjaśnienia zbyt ciemnych wybarwień przez
usunięcie części barwnika z wybarwionego materiału. Nadają się do tego
środki wyrównujące. Środek odbarwiający stosowany jest do usunięcia całego
barwnika z wybarwionej tkaniny. Do
tego celu stosowane są związki redukujące w połączeniu ze środkami wyrównującymi.

582 structural viscosity
Under isothermal and reversible conditions, reduction without hysteresis of
the apparent viscosity under a shearing
load.

fr viscosité structurelle[1]
Dans des conditions isothermes et
réversibles, diminution sans hystérésis
de la viscosité apparente sous l'effet
d'une contrainte de cisaillement croissante.

de Strukturviskosität
Unter isothermen und reversiblen Bedingungen, Verringerung (ohne Hysterese) der scheinbaren Viskosität unter
dem Einfluss einer mechanischen Schubspannung.

el viscosidad estructural (denominación
provisional)
En condiciones isotermas y reversibles,
viscosidad aparente que disminuye sin
histéresis por la acción de un esfuerzo
de cizallamiento creciente.

it viscosità strutturale

ne struktuurviscositeit
Onder isotherme en reversibele kondities,
de verkleining (zonder hysterese) van de
schijnbare viscositeit onder invloed van
een mechanische afschuifspanning.

pl lepkość strukturalna
Obniżenie bez histerezy lepkości pozornej pod wpływem obciążeń ścinających
w warunkach izotermicznych i odwracalnych.

[1] Un autre terme mieux approprié sera ultérieurement choisi pour désigner le concept de l'antidilatance.

583 structure of keratin
fr structure de la kératine
de Keratinstruktur
el estructura de la queratina
it struttura della cheratina
ne struktuur van keratine
pl budowa keratyny

584 styptic pencil
fr crayon hémostatique
de blutstillender Stift
el lápiz hemostático
it matita emostatica
ne bloedstelpende stift
pl laseczka do tamowania krwi przy goleniu

585 substantivity
fr substantivité
de Haftfestigkeit
el substantividad
it sostantività
he substantiviteit
pl substantywność

586 sudoriparous gland, perspiration gland
fr glande sudoripare
de Schweissdrüse
el glándula sudorípara
it ghiandola sudorifera
ne zweetklier
pl gruczoł potowy

587 sulphation
Chemical reaction giving rise to the
formation of a sulphuric ester. In practice,
a mono-sulphuric ester is formed.
fr sulfatation
Réaction chimique permettant d'obtenir

un ester sulfurique. En pratique, on
obtient un mono-ester sulfurique.
de Sulfatierung
Chemische Reaktion, bei welcher ein
Sulfonsäureester entsteht. Im praktischen
Fall entsteht ein Monoester der Schwe-
felsäure.
el sulfatación
Reacción química que permite la obten-
ción de un éster sulfúrico. En la práctica, se obtiene un monoéster sulfúrico.
it solfatazione
Reazione chimica che porta alla formazione di un estere solforico. Generalmente si forma un mono-estere
solforico.
ne sulfatering
Chemische reactie waarbij een zwavel-
zure ester gevormd wordt. In de praktijk
wordt een mono-zwavelzure ester gevormd.
pl siarczanowanie
Reakcja chemiczna prowadząca do utworzenia estrów kwasu siarkowego.
W praktyce powstają zwykle jednoestry
kwasu siarkowego.

588 sulphite addition
Sulphonation brought about by the
reaction of sulphur dioxide or in more
general terms its derivatives (sulphites,
bisulphites) with an electrophilic group.
fr sulfitation
Sulfonation obtenue par réaction de
l'anhydride sulfureux ou plus généralement de ses dérivés (sulfites, bisulfites)
sur un groupement électrophile.
de Sulfitierung
Sulfonierung durch die Reaktion von
Schwefeldioxyd oder deren Derivate
(Sulfite oder Bisulfite) an einer elektro-
philen Gruppe.
el sulfitación
Sulfonación obtenida por reacción del
anhídrido sulfuroso o más generalmente
de sus derivados (sulfitos, bisulfitos)
sobre un grupo electrófilo.
it solfitazione
Reazione chimica di addizione di una
molecola di bisolfito su un legame non
saturo.
ne sulfonering door additie van sulfiet
Een sulfonering door een reactie van
zwaveldioxyde of in meer algemene
termen zijn derivaten (sulfieten, bisul-
fieten) met een elektrofiele groep.

pl addycja siarczku
Sulfonowanie wykonywane poprzez re-
akcję dwutlenku siarki lub jego pochod-
nych (siarczynów lub wodorosiarczy-
nów) z grupą elektrofilową.

589 sulphonation
Chemical reaction leading to the introduction into a molecule of a sulphonic
radical by direct carbon–sulphur linkage.
fr sulfonation
Réaction chimique permettant d'introduire le radical sulfonique dans une
molécule, par liaison directe carbone-
soufre.
de Sulfonierung
Chemische Reaktion, bei welcher der
Sulfonsäurerest in ein Molekül eingeführt wird, mit direkter Bindung von
Schwefel an Kohlenstoff.
el sulfonación
Reacción química que permite introducir
el radical sulfónico en una molécula, por
unión directa carbono–azufre.
it solfonazione
Reazione chimica che introduce il radicale solfonico in una molecola, con legame diretto carbonio–zolfo.
ne sulfonering
Chemische reactie waarbij een sulfon-
zuurgroep via een directe koolstof-
zwavelverbinding in een molecuul wordt
ingevoerd.
pl sulfonowanie
Reakcja chemiczna polegająca na wprowadzeniu do cząsteczki rodnika sul-
fonowego, w której powstaje bezpośrednie wiązanie węgiel–siarka.

590 sun cream
fr crème solaire
de Sonnencreme
el crema solar
it crema solare
ne zonnecrème
pl krem do opalania

591 sun tan preparation
fr préparation pour brunir au soleil
de Bräunungspräparat
el preparado para broncearse al sol
it preparato per abbronzatura
ne middel om bruin te worden
pl preparat wywołujący opaleniznę, samo-
opalacz

592 sunbath
fr bain de soleil

de Sonnenbad
el baño de sol
it bagno di sole
ne zonnebad
pl kąpiel słoneczna

593 sunburn
fr coup de soleil
de Sonnenbrand
el quemaduras del sol
it scottatura da sole
ne zonnebrand
pl rumień słoneczny, zapalenie skóry po naświetlaniu słonecznym, oparzenie słoneczne

594 sunburn cream
fr crème anti-solaire, crème contre les coups de soleil
de Sonnenschutz-Creme
el crema anti-solar, crema para las quemaduras del sol
it crema antisolare, crema contro le scottature da sole
ne anti-zonnebrand-crème
pl krem przeciwsłoneczny

595 sunburn preparation
fr préparation anti-solaire, préparation contre les coups de soleil
de Sonnenschutz-Präparat
el preparado antisolar
it prodotto antisolare
ne middel tegen zonnebrand
pl preparat przeciwsłoneczny

596 sunscreen (agent)
fr écran anti-solaire
de Lichtschutz
el agente protector del sol, filtro antisolar
it filtro antisolare
ne beschermingsmiddel tegen zonnestralen
pl ochrona przeciwsłoneczna, środek przeciwsłoneczny

597 sunscreen preparation
fr préparation anti-solaire
de Lichtschutzmittel
el preparado antisolar
it preparato antisolare
ne zonnebrandmiddel
pl preparat przeciwsłoneczny

598 sunscreening efficiency
fr efficacité contre les coups de soleil
de Lichtschutzwirksamkeit
el eficacia de protección contra el sol
it efficacia contro le scottature

ne beschermende kracht tegen zonnestralen
pl przeciwsłoneczna zdolność ochronna

599 sunscreening product
fr produit protégeant des coups de soleil
de Lichtschutzprodukt
el producto para protegerse del sol
it prodotto capace di proteggere dalle scottature
ne zonnebrandprodukt
pl produkt przeciwsłoneczny

600 superfatting agent
fr agent surgraissant, agent de surgraissage
de Überfettungsmittel
el agente engrasante, agente engrasador
it agente supergrassante
ne overvettingsmiddel
pl środek przetłuszczający

601 superfatting of soap
fr surgraissage
de Überfettung der Seifen
el engrasado del jabón, jabón a la grasa
it aggiunta di supergrassante
ne overvetten
pl przetłuszczanie mydła

602 surface active agent, surfactant, interfacially active agent†
A chemical compound which, when dissolved or dispersed in a liquid, is preferentially adsorbed at an interface, giving rise to a number of physico-chemical or chemical properties of practical interest. The molecule of the compound includes at least one group with an affinity for markedly polar surfaces, ensuring in most cases solubilization in water, and a group which has little affinity for water.
Note: Compositions in practical use are generally mixtures of such compounds.
fr agent de surface
Composé chimique qui, dissous ou dispersé dans un liquide, est préférentiellement adsorbé à une interface, ce qui détermine un ensemble de propriétés physico-chimiques ou chimiques d'intérêt pratique. La molécule du composé comporte au moins un groupement susceptible d'assurer une affinité pour les surfaces nettement polaires, entraînant le plus souvent la solubilisation dans l'eau, et un radical ayant peu d'affinité pour l'eau.
Nota: Les compositions pratiquement utilisées sont en général des mélanges de tels composés.

de grenzflächenaktive Verbindung
Chemische Verbindung, die in einer Flüssigkeit gelöst oder dispergiert an einer Grenzfläche bevorzugt adsorbiert wird; dadurch wird eine Anzahl von praktisch bedeutsamen physikalisch-chemischen oder chemischen Eigenschaften bedingt. Das Molekül dieser Verbindung besitzt wenigstens eine Gruppe, die eine Affinität zu Oberflächen mit einer ausgesprochenen Polarität bewirkt — wodurch im allgemeinen die Löslichkeit in Wasser bedingt wird — sowie einen Rest, der wenig Affinität zu Wasser besitzt.
Anmerkung: Die in der Praxis gebrauchten Produkte sind im allgemeinen Mischungen derartiger Verbindungen.

el agente de superficie, agente tensioactivo
Compuesto químico que, disuelto o dispersado en un líquido, es adsorbido preferentemente en una interfacie, lo que determina la aparición de un conjunto de propiedades físico-químicas de importancia práctica. La molécula de estos compuestos contiene al menos un grupo con afinidad para superficies francamente polares, lo que produce en general la solubilización en agua del compuesto, y un radical con poca afinidad para el agua.
Observación: Las composiciones que se utilizan en la práctica son generalmente mezclas de tales sustancias.

it tensioattivo
Composto chimico che disciolto o disperso in un liquido è preferenzialmente assorbito a una interfaccia, ciò che determina un insieme di proprietà fisicochimiche o chimiche di interesse pratico. La molecola del composto comprende almeno un gruppo con affinità per le superfici nettamente polari, in grado di assicurare il più delle volte la dissoluzione in acqua e un radicale avente poca affinità per l'acqua.
Nota: I prodotti di uso comune sono generalmente miscele di questi composti.

ne oppervlakaktieve stof
Een chemische verbinding die bij oplossing of dispergering in een vloeistof de neiging vertoont aan een grensvlak geadsorbeerd te worden met als gevolg een aantal fysisch-chemische of chemische eigenschappen van praktisch belang. Het molecuul van de verbinding bevat ten minste één groep met een affiniteit voor uitgesproken polaire vlakken, die de verbinding in de meeste gevallen in water oplosbaar maakt, alsmede een groep die een geringe affiniteit voor water heeft.
Opmerking: De in praktijk gebruikte samenstellingen zijn in het algemeen mengsels van zulke verbindingen.

pl środek powierzchniowo czynny
Związek chemiczny ulegający w sposób uprzywilejowany adsorbcji na granicy faz po rozpuszczeniu lub zdyspergowaniu w cieczy, co pociąga za sobą wystąpienie szeregu własności fizyko-chemicznych lub chemicznych mających praktyczne znaczenie. Cząsteczka takiego związku posiada co najmniej jedną grupę wykazującą powinowactwo do wyraźnie polarnych powierzchni, zapewniającą w większości przypadków jego rozpuszczalność w wodzie, oraz grupę o małym powinowactwie do wody.
Uwaga: Produkty stosowane w praktyce są na ogół mieszaninami takich związków.
† See appendix

603 surface active derivate
Product endowed with surface activity.
fr dérivé tensio-actif
Produit doué de tensio-activité.
de Tensid
Grenzflächenaktive Verbindung.
Anmerkung: Die Bezeichnung Tensid wird in der deutschen Fachsprache als Synonym für den Begriff Nr. 602 benutzt.
el tensioactivo
Producto dotado de tensioactividad.
it derivato tensioattivo
ne grensvlakaktieve stof
Chemische verbinding met grensvlakaktieve eigenschappen.
pl związek powierzchniowo czynny
Związek obdarzony własnościami powierzchniowo czynnymi.

604 surface activity
All the properties particular to surface active agents in solution and operating at the interfaces.
fr activité de surface
Ensemble des propriétés particulières aux agents de surface en solution et s'exerçant aux interfaces.
de Grenzflächenaktivität
Gesamtheit der den Tensiden zugehörigen Eigenschaften, welche an Phasengrenzen in Erscheinung treten.

el actividad de superficie, actividad superficial, tensioactividad
Conjunto de propiedades especiales de
los agentes de superficie en disolución
que se ejercen en las interfacies.

it attività superficiale
Insieme delle proprietà caratteristiche dei
tensioattivi in soluzione, che si manifestano alle interfacce.

ne oppervlakaktiviteit
Alle speciale eigenschappen van oppervlakaktieve stoffen in oplossing en
werkzaam aan de grensvlakken.

pl aktywność powierzchniowa
Ogół własności charakterystycznych dla
rozpuszczonych związków powierzchniowo czynnych działających na granicach faz.

605 surface forces
fr forces de surface
de Oberflächenkräfte
el fuerzas de superficie
it forze di superficie
ne oppervlaktekrachten
pl siły powierzchniowe

606 surface phenomena
Phenomena the effects of which (mechanical, electrical, optical, etc.) become
apparent at the surface separating two
phases (liquid–gas, liquid–solid, liquid–
liquid, or gas–solid).

fr phénomène de surface
Phénomène dont les effets (mécaniques,
électriques, optiques, etc.) se manifestent
à la surface de séparation de deux phases
(liquide–gaz, liquide–solide, liquide–liquide ou gaz–solide).

de Grenzflächenerscheinungen
Erscheinungen deren mechanische, elektrische, optische Effekte, usw. sich an der
Grenzfläche zweier Phasen (flüssig–gasförmig, flüssig–fest, flüssig–flüssig oder
gasförmig–fest) bemerkbar machen.

el fenómenos de superficie
Fenómenos cuyos efectos mecánicos,
eléctricos, ópticos, se manifiestan en
la superficie de separación de dos fases
(líquido–gas, líquido–sólido, líquido–
líquido o gas–sólido).

it fenomeni di superficie
Fenomeni i cui effetti (meccanici, elettrici,
ottici, ecc.) si manifestano alla superficie
di separazione di due fasi (liquido–gas,
liquido–solido, liquido–liquido o gas–
solido).

ne oppervlakverschijnselen
Verschijnselen waarvan de resultaten
(mechanische, elektrische, optische enz.)
merkbaar zijn aan het scheidingsvlak
tussen twee fasen (vloeistof–gas, vloeistof–vaste stof, vloeistof–vloeistof, of
gas–vaste stof).

pl zjawiska powierzchniowe
Zjawiska, których efekty (mechaniczne,
elektryczne, optyczne itd.) uwidaczniają
się na powierzchni rozdzielającej dwie
fazy (ciecz–gaz, ciecz–ciało stałe, ciecz–
ciecz lub gaz–ciało stałe).

607 surface tension (symbols γ and γ_s)
The force per unit of length resulting
from the free surface energy. It is expressed in newtons per metre $(N/m)^{1)}$.

fr tension superficielle (symboles γ et γ_s)
Force par unité de longueur, résultante
de l'énergie superficielle libre. Elle
s'exprime en newtons par mètre $(N/m)^{1)}$.

de Oberflächenspannung (Symbole γ und γ_s)
Kraft pro Längeneinheit, die aus der
freien Oberflächenenergie resultiert. Sie
wird in Newton pro Meter (N/m)
ausgedrückt$^{1)}$.

el tensión superficial
Fuerza por unidad de longitud, resultante de la energía superficial libre.
Numéricamente es igual a la energía
superficial por unidad de superficie.

it tensione superficiale
Forza per unità di lunghezza, risultante
dall'energia superficiale libera.
E'numericamente uguale all'energia superficiale libera per unità di superficie
e si esprime in dine per centimetro$^{2)}$.

ne oppervlakspanning
De kracht per lengte-eenheid, voortkomende uit de vrije oppervlakenergie.
Deze wordt uitgedrukt in Newtons per
meter $(N/m)^{3)}$.

pl napięcie powierzchniowe (symbole γ i γ_s)
Siła na jednostkę długości wynikająca
ze swobodnej energii powierzchniowej.
Siła ta wyrażana jest w niutonach na
metr N/m.

[1] 1 N/m = 10^3 dynes/cm.
[2] La dine per centimetro (din/cm) è l'unità del
sistema CGS. L'unità SI è il Newton per metro
(N/m): 1 N/m = 10^3 din/cm.
[3] 1 N/m (newton per meter) = 10^3 dynes/cm.

608 suspending power
In the case of solutions of surface active
agents: the ability of certain substances

to maintain in suspension particles insoluble in the solution.

Note: The suspending power can vary very considerably depending upon the nature of these particles.

fr pouvoir suspensif
Dans le cas de solutions d'agents de surface: degré d'aptitude de certaines substances à maintenir en suspension des particules insolubles dans la solution.

Nota: Le pouvoir suspensif peut varier très fortement selon la nature de ces particules.

de Suspendiervermögen
Fähigkeit gewisser Substanzen, ungelöste Partikel in Suspension zu halten.

Anmerkung: Das Suspendiervermögen ist ausserordentlich spezifisch für die Art der suspendierten Substanz.

el poder de suspensión
En el caso de disoluciones de agentes de superficie: capacidad de determinadas substancias para mantener en suspensión partículas sólidas insolubles en la disolución.

Observación: El poder de suspensión puede variar considerablemente según la naturaleza de las partículas.

it potere sospendente
Grado dell'attitudine di certe sostanze a mantenere in sospensione particelle insolubili nella soluzione.

Nota: Il potere sospendente può variare notevolmente in funzione della natura delle particelle.

ne suspenderend vermogen
In het geval van oplossingen van oppervlakaktieve stoffen: het vermogen van bepaalde stoffen om deeltjes die onoplosbaar in de oplossing zijn in suspensie te houden.

Opmerking: Het suspenderend vermogen kan sterk variëren, afhankelijk van de aard van deze deeltjes.

pl zdolność utrzymywania w zawiesinie
W przypadku roztworów związków powierzchniowo czynnych zdolność pewnych substancji do utrzymywania w zawiesinie cząstek nierozpuszczalnych w roztworze.

Uwaga: Zdolność utrzymywania w zawiesinie może różnić się znacznie w zależności od charakteru zawieszonych cząstek.

609 sweat
fr sueur
de Schweiss
el sudor
it sudore
ne zweet
pl pot

610 sweat secretion
fr sécrétion sudorale
de Schweissabsonderung
el secreción de sudor
it secrezione di sudore
ne zweetafscheiding
pl wydzielanie potu

611 sweating, sudation, perspiration
fr sudation, transpiration
de Schwitzen
el transpiración, sudor
it traspirazione
ne transpiratie
pl pocenie się

612 sweating of soap
fr ressuage des savons
de Schwitzen der Seifen
el transpiración del jabón, efflorescencia del jabón
it trasudamento del sapone
ne zweten van de zeep
pl pocenie się mydła

613 swelling
fr gonflement
de Quellung
el ahuecado, hinchado
it gonfiore, rigonfiamento
ne zwellen
pl pęcznienie

614 swelling of the skin
fr gonflement de la peau
de Quellung der Haut
el hinchazón de la piel
it rigonfiamento della pelle
ne zwellen van de huid
pl pęcznienie skóry, obrzmienie skóry

615 synergy, synergistic effect
A mixture, in given proportions, of two surface active agents or other chemicals, within certains limits of their respective concentration, exhibits a given effectiveness for a given concentration of the mixture in the medium in which the measurement is carried out. There is synergy when this concentration is lower than that which would result from the linear combination, in the same proportions, of the concentrations which, for each constituent considered separately, would be necessary to achieve the same effectiveness. The above constituents referred to may themselves be mixtures.

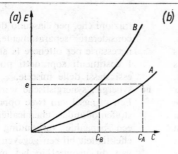

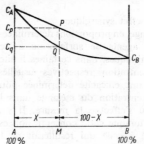

en Fig. (a) — The curves show the efficiencies E of the constituents A and B considered separately (as ordinate) as a function of their concentrations (as abscissae). For a given efficiency e there must be either a concentration C_A for pure A or a concentration C_B for pure B.
Fig. (b) — With the composition of mixtures of the constituents A and B shown on the abscissa, a point M corresponds to a mixture of $x\%$ of B and $(100-x)\%$ of A. The concentrations C_M in the testing medium, of mixtures with a given efficiency e are shown as ordinate. If there is a linear relation between the efficiency e and x, the concentration of M giving the efficiency e will be equal to C_p, P being on the straight line $C_A C_B$. There is synergy between A and B if the concentration C_q of M required to achieve the efficiency e is lower than C_p.

fr Fig. (a) — Les courbes donnent en ordonnée les efficacités E des constituants A et B considérés isolément, en fonction de leur concentration portée en abscisse. Pour une efficacité donnée e, il faut avoir soit une concentration C_A en A pur, soit une concentration C_B en B pur.
Fig. (b) — La composition des mélanges des constituants A et B étant portée en abscisse, un point M correspond à un mélange de $x\%$ de B et de $(100-x)\%$ de A. En ordonnée, sont indiquées les concentrations C_M dans le milieu où s'effectue la mesure des mélanges de composition M donnant une efficacité donnée e. S'il existe une relation linéaire entre l'efficacité e et x, la concentration de M donnant l'efficacité e doit être égale à C_p, P étant sur la ligne droite $C_A C_B$. Il y a synergie entre A et B si, en réalité, la concentration C_q du mélange M pour obtenir cette efficacité e est inférieure à C_p.

de Abb. (a) — Die Kurven zeigen als Ordinate die Wirkung E der Bestandteile A und B für sich betrachtet in Abhängigkeit von ihrer auf die Abszisse eingetragenen Konzentration. Für eine gegebene Wirkung e braucht man eine Konzentration C_A an reinem A, sowie eine Konzentration C_B an reinem B.
Abb. (b) — Die Zusammensetzung der Mischung der Bestandteile A und B ist auf der Abszisse aufgetragen. Ein Punkt M entspricht einer Mischung von $x\%$ B und $(100-x)\%$ A. Auf der Ordinate sind die Konzentrationen C_M in der Lösung angegeben, bei denen das Mischungsverhältnis von der Zusammensetzung M eine gegebene Wirkung e erzielt. Nach der Mischungsregel ist die für die Wirkung e notwendige Konzentration M gleich C_p, entsprechend dem Punkt P auf der Geraden $C_A C_B$. Ein synergistischer Effekt besteht zwischen A und B, wenn die zur Erzielung der Wirkung e notwendige Konzentration C_q der Mischung M in Wirklichkeit niedriger ist als C_p.

el Fig. (a) — Las curvas dan en ordenadas las eficacias E de los constituentes A y B aislados, en función de su concentración en abscisas. Para una eficacia dada e, hace falta una concentración C_A de A puro o una concentración C_B de B puro.
Fig. (b) — La composición de la mezcla de los constituyentes A y B está dada en abscisas. Un punto M corresponde a una mezcla de $x\%$ de B y de $(100-x)\%$ de A. En ordenadas se indican las concentraciones C_M de las mezclas de composición M para una eficacia determinada e. Según la regla de mezclas, la concentración de M que da la eficacia e debería ser igual a C_p siendo P un punto de la recta $C_A C_B$. Existe sinergia entre A y B si, en realidad, la concentración C_q de la mezcla M, para la eficacia e, es inferior a C_p.

it Fig. (a) — Le curve indicano l'attività E (nelle ordinate) dei costituenti A e B, considerati isolatamente, in funzione della loro concentrazione (sulle ascisse). Per una data attività e sarà necessaria una concentrazione di A puro C_A oppure una concentrazione C_B di B puro.
Fig. (b) — Avendo riportato sulle ascisse la composizione delle miscele dei costituenti A e B, un punto M corrisponde ad una miscela di $x\%$ di B e $(100-x)\%$ di A. Sulle ordinate sono indicate le concentrazioni C_M, nel mezzo di prova, di miscele di composizione M aventi una attività e. Se vi è relazione lineare fra l'attività e ed x, la concentrazione di M avente l'attività e sarà uguale a C_p, con P situata sulla linea retta $C_A C_B$. Vi è sinergia fra A e B se la concentrazione C_q della miscela M necessaria per ottenere l'attività e è inferiore a C_p.

ne Fig. (a) — De krommen geven de efficiënties E van de afzonderlijke bestanddelen A en B (ordinaten) als functie van hun concentraties (abscissen) weer. Bij een gegeven efficiëntie e is er òf een concentratie C_A voor zuiver A òf een concentratie C_B voor zuiver B.
Fig. (b) — Bij de samenstelling van mengsels van de bestanddelen A en B als weergegeven op de abscis korrespondeert een zeker punt M met een mengsel van $x\%$ B en $(100-x)\%$ A. De concentraties C_M van mengsels met een gegeven efficiëntie e in het te onderzoeken milieu zijn weergegeven in de vorm van ordinaten. De wet van de mengsels vereist dat de concentratie van M die de efficiëntie e oplevert gelijk is aan C_p, waarbij P gelegen is op de rechte lijn $C_A - C_B$. Er bestaat synergie tussen A en B wanneer de concentratie C_q van M die vereist is om de efficiëntie e te bereiken, kleiner is dan C_p.

pl Rys. (a) — Krzywe przedstawiają efekt E składników A i B rozpatrywanych oddzielnie w funkcji ich stężenia. Dla uzyskania danego efektu e wymagane jest stężenie C_A dla czystego związku A lub stężenie C_B dla czystego związku B.
Rys. (b) — Skład mieszaniny składników A i B podany jest na osi odciętych. Punkt M odpowiada mieszaninie zawierającej $x\%$ składnika B i $100-x\%$ składnika A. Na osi rzędnych podane jest stężenie C_M występujące w badanym roztworze, przy którym mieszanina osiąga efekt e. Jeśli istnieje liniowa zależność pomiędzy efektem e a x, to stężenie M, przy którym osiągany jest efekt e. będzie równe stężeniu C_p odpowiadającemu punktowi P na prostej $C_A C_B$. Zjawisko synergizmu pomiędzy substancjami A i B występuje gdy stężenie C_q mieszaniny M wymagane dla uzyskania efektu e jest niższe niż stężenie C_p.

fr synergie, effet synergique
Un mélange, en proportions déterminées, de deux agents de surface ou autres agents chimiques, dans certaines limites de concentration respectives usuelles, présente une efficacité déterminée pour une concentration du mélange dans le milieu où s'effectue la mesure. Il y a synergie lorsque cette concentration est inférieure à celle qui résulterait de la combinaison linéaire dans les mêmes proportions, des concentrations qui, pour chacun des constituants considérés isolément, seraient nécessaires à l'obtention de la même efficacité. Les constituants précédents peuvent être eux-mêmes des mélanges.

de Synergismus, synergistischer Effekt
Effekt durch den eine Mischung aus zwei Tensiden oder anderen Substanzen innerhalb gewisser, wechselseitig bedingter Konzentrationsgrenzen in der Lösung eine bestimmte Wirkung schon bei einer kleineren Konzentration erreicht, als einer linearen Abhängigkeit der für diese Wirkung notwendigen Konzentration vom Mischungsverhältnis entspricht. Die genannten Substanzen können ihrerseits wieder Mischungen sein.

el sinergia, efecto sinérgico
Una mezcla de proporciones determinadas de dos agentes de superficie o de otras substancias, entre ciertos límites de sus concentraciones respectivas, posee una determinada eficacia a una concentración dada. Existe sinergia si esta concentración es inferior a la que resultaría de la combinación lineal, en las mismas proporciones, de las concentraciones de cada uno de los constituyentes puros, que serían necesarias para alcanzar la mencionada eficacia. Los constituyentes pueden ser a su vez mezclas.

it sinergia
Una miscela in determinate proporzioni di due tensioattivi o di altre sostanze chimiche, entro certi limiti delle rispettive concentrazioni, presenta una data attività per una data concentrazione della miscela nel mezzo nel quale si effettua la misura. Si ha sinergia quando questa concentrazione e inferiore a quella che risulterebbe dalla combinazione lineare nelle stesse proporzioni, delle concen-trazioni che, per ciascuno dei costituenti considerati separatamente, sarebbero necessarie per ottenere la stessa attività. I costituenti sopradetti possono essere essi stessi delle miscele.

ne synergie, synergistische werking
Een mengsel van twee oppervlakaktieve stoffen of andere chemicaliën vertoont in een bepaalde verhouding een zekere effektiviteit bij een gegeven concentratie van dit mengsel in het milieu waarin de bepaling wordt verricht. Er is sprake van synergie wanneer deze concentratie lager is dan die welke ontstaan uit de lineaire kombinatie, in dezelfde verhouding, van de concentraties die voor elk bestanddeel afzonderlijk nodig zouden zijn om dezelfde effektiviteit te bereiken. De bovengenoemde bestanddelen kunnen zelf mengsels zijn.

pl synergizm, efekt synergetyczny
Mieszanina w określonym stosunku dwóch związków powierzchniowo czynnych lub innych związków chemicznych w pewnych zakresach ich stężeń wykazuje określone działanie przy danym stężeniu mieszaniny w środowisku, w którym wykonywany jest pomiar. Zjawisko synergizmu występuje wtedy, gdy stężenie to jest niższe niż wynikałoby z zależności liniowej określającej, rozpatrywane oddzielnie, ilości składników mieszaniny potrzebne dla uzyskania tego samego działania. Składniki mieszaniny same mogą być również mieszaninami.

616 synthetic bar
fr saponide façonné, détergent en pain, savonnette synthétique, barre synthé-tique, barre détergente, syndet en mor-ceaux
de synthetische Stückseife
el detergente en barra, detergente en pas-tilla
it detergente sintetico in barre, saponetta sintetica
ne stuk synthetische zeep
pl mydło syntetyczne w kawałku

617 synthetic shaped washing agent
fr syndet moulé
de synthetisches geformtes Waschmittel
el detergente sintético compuesto
it detergente sintetico in pezzi
ne gevormd synthetisch wasmiddel
pl syntetyczny środek piorący w kawałku

T

618 tan
fr brunissage, hâle
de Sonnenbräune
it abbronzatura
el atezamiento, tostado del sol
ne bruin zijn
pl opalenizna

619 tearing strain, tear strength
fr résistance à la déchirure
de Reissfestigkeit
el resistencia a la rotura, resistencia al rasgado
it resistenza alla rottura
ne scheursterkte
pl wytrzymałość na rozerwanie, wytrzymałość na rozdzieranie

620 teeth disorder
fr troubles dentaires
de schlechte Zähne
el trastornos dentales
it malattie dei denti
ne tandaandoening
pl schorzenie zębów, zepsute zęby

621 temperature of clarification
In the case of non-ionic surface active agents which exhibit a cloud temperature: the temperature at which the mixture of the two liquid phases becomes homogeneous on cooling.
Note: The temperature of clarification is often determined as "cloud point".
fr température de clarification
Dans le cas d'agents de surface non-ioniques présentant une température de trouble: température à partir de laquelle le mélange des deux phases liquides est devenu homogène par refroidissement.
Nota: La température de clarification est souvent déterminée comme "point de trouble".
de Klarpunkt (oft als Trübungspunkt bezeichnet)
Im Falle gewisser nicht-ionischer Tenside, welche einen Trübungspunkt haben: Temperatur, unterhalb welcher die Mischung der beiden flüssigen Phasen durch Abkühlung homogen geworden ist.
el temperatura .de clarificación
En el caso de agentes de superficie no iónicos que posean una temperatura de enturbiamiento: temperatura a partir de la cual la mezcla de las dos fases

líquidas se hace homogénea por enfriamiento.
Observación: La temperatura de clarificación se determina frecuentemente "punto de enturbiamiento".
it temperatura di chiarificazione
Nel caso dei tensioattivi non-ionici che presentano una temperatura di intorbidamento: temperatura, alla quale la miscela delle due fasi liquide diventa omogenea per raffréddamento.
Nota: La temperatura di chiarificazione è sovente determinata come "temperatura di intorbidamento".
ne ophelderingspunt
In het geval van niet-ionogene oppervlakaktieve stoffen waarbij het troebelpunt overschreden is; bij afkoelen de temperatuur waarbij het mengsel van de twee vloeibare fasen homogeen wordt.
pl temperatura klarowania
W przypadku niejonowych związków powierzchniowo czynnych posiadających temperaturę zmętnienia jest to temperatura, w której mieszanina dwóch faz ciekłych staje się jednorodna przy oziębianiu.
Uwaga: Temperatura klarowania często nazywana jest punktem zmętnienia.

622 temporary hardness
fr dureté temporaire
de vorübergehende Härte
el dureza temporal
it durezza temporanea
ne tijdelijke hardheid
pl twardość przemijająca (wɔdy)

623 tensile strength
fr résistance
de Festigkeit
el resistencia, fuerza de la fibra
it resistenza alla trazione
ne sterkte
pl wytrzymałość na rozciąganie

624 tepid wave
fr permanente tiède
de Mildwelle, Lauwelle
el permanente tibia, permanente tempada
it permanente tiepide
ne lauw onduleren
pl ondulacja na ciepło

625 test swatches of soiled fabric
fr éprouvette de tissu sali standard

de angeschmutzter Teststreifen
el pruebas de tejidos ensuciados
it campioni standard di tessuto sporcato
ne gestandariseerd bevuild proeflapje
pl testowe kawałki zabrudzonej tkaniny

626 textile cleanser
fr détersif pour les textiles
de Textilreinigungsmittel
el detergente para los tejidos, detergente para la industria textil
it detergente per tessili
ne textielwasmiddel
pl środek piorący do wyrobów włókienniczych

627 therapeutic power
fr pouvoir thérapeutique
de therapeutische Kraft
el poder terapéutico
it potere terapeutico
ne helende kracht
pl zdolność lecznicza

628 thickening capacity
fr capacité d'épaississement, pouvoir épaississant
de Verdickungs-Kapazität
el capacidad de espesar, poder espesante
it potere ispessente
ne verdikkend vermogen
pl zdolność zagęszczania

629 thioglycerol
fr thioglycérol
de Thioglycerin
el tioglicerol
it tioglicerolo
ne thioglycerol
pl tiogliceryna

630 thioglycolic acid
fr acide thioglycolique
de Thioglykolsäure
el ácido tioglicólico
it acido tioglicolico
ne thioglycolzuur
pl kwas tioglikolowy

631 thiolactic acid
fr acide thiolactique
de Thiomilchsäure
el ácido tioláctico (que elimina el colorante)
it acido tiolattico
ne thiomelkzuur
pl kwas tiomlekowy

632 thixotropy
Under isothermal and reversible conditions, reduction with hysteresis of the apparent viscosity under a shearing load.
fr thixotropie
Dans des conditions isothermes et réversibles, diminution avec hystérésis de la viscosité apparente sous l'effet d'une contrainte de cisaillement croissante.
de Thixotropie
Unter isothermen und reversiblen Bedingungen, Verringerung (mit Hysterese) der scheinbaren Viskosität unter dem Einfluss einer mechanischen Schubspannung.
el tixotropía
En condiciones isotermas y reversibles, disminución con histéresis de la viscosidad aparente, por la acción de un esfuerzo de cizallamiento creciente.
it tixotropia
ne tixotropie
Onder isotherme en reversibele omstandigheden verlaging met hysterese van de schijnbare viscositeit onder de invloed van een mechanische afschuifspanning.
pl tiksotropia
Obniżenie (z histerezą) pozornej lepkości pod wpływem obciążeń ścinających w warunkach izotermicznych i odwracalnych.

633 time of wetting
fr durée de mouillage, temps de mouillage
de Benetzungszeit
el duración de la humectación, tiempo de humectación
it durata dell'ammollo, tempi di ammollo
ne bevochtigingstijd
pl czas zwilżania

634 tinted foundation cream
fr crème de base teintée
de getönte Grundierungscreme
el crema de base teñida
it crema di base per tintura
ne gekleurde basiscrème
pl tonizujący krem pod puder

635 toilet soap
Perfumed soap for toilet requirements, containing 76–83% fatty acids and certain types of superfatting additives, cosmetics, etc.
fr savon de toilette
Savon parfumé employé pour les soins de toilette, contenant de 76 à 83% d'acides gras et aussi parfois d'additifs surgraissants, cosmétiques et autres.

de Feinseife (Toiletteseife)
Ca. 76–83% Fettsäure enthaltende parfumierte Seife, die zur Körperpflege dient, auch mit überfettenden, hautpflegenden, kosmetischen oder sonstigen allgemein hautschützenden Zusätzen.

el jabón de focador
Jabón perfumado de 76 a 83% de ácidos grasos, empleado para los cuidados corporales, puede contener aditivos sobreengrasantes, cosméticos u otros.

it sapone da toeletta
Sapone profumato per l'igiene personale contenente acidi grassi nella misura dal 76 all' 83% oltre ad additivi surgrassanti, cosmetici, ecc.

ne toiletzeep
Ca. 76–83% vetzuur bevattende, geparfumeerde zeep, die voor de lichaamsverzorging dient en die soms kosmetische toevoegingen bevat als b.v. overvettingsmiddelen.

pl mydło toaletowe
Perfumowane mydło, przeznaczone do celów toaletowych zawierające 76–83% kwasów tłuszczowych oraz dodatki przetłuszczające, kosmetyczne itd.

636 tonic beauty lotion
fr lotion de beauté tonique
de flüssiges Schönheitstonikum
el loción tónica de belleza
it tonico di bellezza
ne opwekkend schoonheidswater
pl tonizujący płyn upiększający

637 tooth brush
fr brosse à dents
de Zahnbürste
el capillo de dientes
it spazzolino da denti
ne tandenborstel
pl szczotka do zębów

638 tooth enamel
fr émail de la dent
de Zahnschmelz
el esmalte del diente
it smalto del dente
ne tandemail
pl emalia nazębna

639 toothpaste
fr pâte dentifrice

de Zahnpaste
el dentífrico, pasta de dientes
it pasta dentifricia
ne tandpasta
pl pasta do zębów

640 toothpowder
fr poudre dentifrice
de Zahnpulver
el polvo dentífrico
it polvere dentifricia
ne tandpoeder
pl proszek do zębów

641 topical anaesthetic
fr anesthésique local
de lokales Betäubungsmittel
el anestésico local
it anestetico locale
ne middel voor plaatselijke verdoving
pl środek do znieczulania miejscowego

642 tragacanth
fr gomme adragante
de Traganth
el goma tragacanto, goma adragante
it gomma adragante
ne tragant
pl tragakant, guma tragakantowa

643 treatment cream
fr crème de traitement
de Kurcreme
el crema de tratamiento
it crema da trattamento
ne behandelingscrème
pl krem leczniczy

644 turbidity measurement
fr mesure de turbidité
de Trübungsmessung
el medida de la turbidez
it misura d'intorbidamento
ne troebelings-meting
pl pomiar zmętnienia

645 turbidity point, cloud point
fr point de trouble
de Trübungspunkt
el punto de enturbiamento
it punto di intorbidamento
ne troebelingspunt
pl temperatura zmętnienia, punkt zmętnienia

U

646 unsaponifiable matter
The whole of the constituents, soluble in fatty matter and insoluble in water, which cannot be modified by a reaction of saponification producing a salt.

Note: In practice and for analytical determination: the products present in the substance analysed which, after saponification of the latter with an alkaline hydroxide and extraction by a specified solvent, remain non-volatile under the defined conditions of test.

fr insaponifiable
Ensemble des constituants solubles dans la matière grasse et insolubles dans l'eau, qui ne sont pas susceptibles d'être modifiés par la réaction de saponification en donnant un sel.

Nota: En pratique, et pour les déterminations analytiques: ensemble des produits présents dans la substance analysée qui, après saponification de celle-ci par un hydroxyde alcalin et extraction par un solvant spécifié, restent non volatils dans les conditions opératoires décrites.

de Unverseifbares
Gesamtheit der fettlöslichen und wasser-unlöslichen organischen Substanzen, die durch die Verseifungsreaktion nicht in Salze übergeführt werden können.

Anmerkung: In der Praxis und bei analytischen Bestimmungen: Gesamtheit der in der analysierten Substanz vorhandenen Produkte, welche nach der Verseifung durch Alkali und Extraktion durch ein geeignetes Lösungsmittel unter den üblichen Arbeitsbedingungen als nicht flüchtiger Rückstand verbleiben.

el insaponificable
Conjunto de constituyentes solubles en la materia grasa e insolubles en agua que no son susceptibles de ser modificados por la reacción de saponificación para formar una sal.

Observación: En la práctica, así como para determinaciones analíticas: conjunto de productos presentes en la substancia analizada que, después de saponificación de ésta por un hidróxido alcalino y de extracción por un disolvente especificado, quedan sin volatilizarse en las condiciones operativas prescritas.

it —

ne niet-verzeepbare substantie
Het geheel van produkten, oplosbaar in vetten en onoplosbaar in water, die door een verzepingsreactie niet kunnen worden omgezet in een zout.

pl substancje niezmydlające się
Ogół składników rozpuszczalnych w sub-stancji tłuszczowej a nierozpuszczalnych w wodzie, które nie mogą zostać prze-prowadzone w sole w reakcji zmydlania.

Uwaga: W praktyce jak również przy oznaczeniach analitycznych jest to ogół produktów wystę-pujących w analizowanej substancji, które po wykonaniu zmydlenia przy pomocy alkaliów i ekstrakcji odpowiednim rozpuszczalnikiem po-zostają w normalnych warunkach prowadzenia analizy jako nielotna pozostałość.

647 unsaponified matter
A saponifiable substance which has survived a saponification reaction.

fr insaponifié, non saponifié
Substance saponifiable ayant échappé à la réaction de saponification.

de Unverseiftes, Nichtverseiftes
Verseifbare Substanz, die durch die Verseifungsreaktion nicht erfasst worden ist.

el insaponificado, no saponificado
Substancia saponificable que ha escapado a la reacción de saponificación.

ne niet-verzeepte substantie
Een verzeepbare substantie die bij de verzepingsbehandeling niet verzeept is.

it —

pl substancje niezmydlone
Substancje ulegające normalnie zmydle-niu, które nieprzereagowały podczas reakcji zmydlenia.

648 unsulphatable matter
A constituent which is not capable of undergoing a sulphation reaction.

fr insulfatable
Constituant qui n'est pas susceptible de subir une réaction de sulfatation.

de Unsulfatierbares
Bestandteil, der nicht sulfatiert werden kann.

el insulfatable
Constituyente que no es susceptible de experimentar una reacción de sulfatación.

ne niet-sulfateerbare substantie
Een substantie die bij de sulfateringsbe-handeling niet reageert.

pl substancje niesiarczanujące się
Składniki, które nie są zdolne do ulega-nia reakcji siarczanowania.

649 unsulphated matter
A sulphatable substance which has sur-vived a sulphation reaction and/or is converted into an unsulphatable product during this reaction.

fr insulfaté, non sulfaté
Substance sulfatable ayant échappé à la réaction de sulfatation, et/ou s'étant transformée en produit non sulfatable au cours de celle-ci.

de Unsulfatiertes, Nichtsulfatiertes
Sulfatierbare Substanz, die durch eine Sulfatierungsreaktion nicht sulfatiert worden ist, und/oder im Verlauf derselben in ein nicht sulfatierbares Produkt übergeführt worden ist.

el insulfatado, no sulfatado
Substancia sulfatable que ha escapado a la reacción de sulfatación y/o se ha transformado en producto no sulfatable en el curso de dicha reacción.

it —

ne niet-gesulfateerde substantie
Een sulfateerbare substantie die bij de sulfateringsbehandeling niet gesulfateerd is of is omgezet in een niet-sulfateerbare substantie.

pl substancje niezsiarczanowane
Substancje siarczanujące się, które nie uległy reakcji siarczanowania i/lub zostały przeprowadzone podczas tej reakcji w substancje niesiarczanujące się.

650 unsulphonatable matter
A constituent which is not capable of undergoing a sulphonation reaction.

fr insulfonable
Constituant qui n'est pas susceptible de subir une réaction de sulfonation.

de Unsulfonierbares
Bestandteil, der nicht sulfoniert werden kann.

el insulfonable
Constituyente que no es susceptible de experimentar una sulfonación.

it —

ne niet-sulfoneerbare substantie
Een substantie die bij de sulfoneringsreactie niet reageert.

pl substancje niesulfonujące się
Składniki, które nie są zdolne do ulegania reakcji sulfonowania.

651 unsulphonated matter
A sulphonatable substance which has survived a sulphonation reaction and/or is converted into an unsulphonatable product during this reaction.

fr insulfoné, non sulfoné
Substance sulfonable ayant échappé à la réaction de sulfonation et/ou s'étant transformée en produit non sulfonable au cours de celle-ci.

de Unsulfoniertes, Nichtsulfoniertes
Sulfonierbare Substanz, die im Laufe einer Sulfonierungsreaktion nicht sulfoniert worden ist und/oder im Verlauf derselben in ein nicht sulfonierbares Produkt umgewandelt worden ist.

el insulfonado, no sulfonado
Substancia sulfonable que ha escapado a la reacción de sulfonación y/o se ha transformado en producto no sulfonable en el curso de dicha reacción.

it —

ne niet-gesulfoneerde substantie
Een sulfoneerbare substantie die bij de sulfoneringsreactie niet gesulfoneerd is of is omgezet in een niet-sulfoneerbare substantie.

pl substancje niezsulfonowane
Substancje sulfonujące się, które nie uległy reakcji sulfonowania i/lub zostały przeprowadzone podczas tej reakcji w substancje niesulfonujące się.

652 untreated skin
fr peau non traitée
de unbehandelte Haut
el piel no tratada
it pelle non trattata
ne onbehandelde huid
pl skóra nietraktowana, skóra nieleczona

653 used detergent solution
fr solution de lavage usée
de Schmutzflotte, gebrauchte Waschlauge, schmutzbelastete Waschlauge
el solución detergente usada
it bagno detergente usato
ne gebruikte wasvloeistof
pl zużyta kąpiel piorąca

V W

654 vanishing cream
fr vanishing cream
de Vanishingcreme
el vanishing cream
it crema evanescente
ne dagcrème
pl krem półtłusty

655 vegetable dye
fr colorant végétal
de vegetabilisches Haarfärbemittel
el colorante vegetal
it colorante vegetale
ne plantekleurstof
pl barwnik roślinny

656 wash liquor, detergent solution, washing bath
fr liqueur de lavage, bain de lavage, solution détergente
de Flotte, Waschlauge
el líquido del lavado, baño del lavado, solución detergente
it bagno detergente, bagno di lavaggio, soluzione di lavaggio
ne wasvloeistof
pl kąpiel piorąca

657 washable
fr lavable, lessivable
de waschbar
el se puede lavar, lavable
it lavabile
ne wasbaar
pl nadający się do prania

658 washing
The operation, by means of aqueous solutions of electrolytes, leading to completion of the separation of the glycerine and the impurities from the curd soap arising from the graining out.
fr lavage
Opération permettant, au moyen de solutions aqueuses d'électrolytes, de compléter la séparation de la glycérine et des impuretés du savon grainé venant du relargage.
de Waschen
Vorgang der Reinigung des geronnenen Seifenleims mittels wässriger Elektrolyt-Lösungen, mit dem Ziel, die Abscheidung von Glycerinresten und von dispergierbaren Verunreinigungen zu vervollständigen.
el lavado
Depuración del jabón graneado por mezcla intensiva con disoluciones acuosas de electrólitos (en general, de cloruro sódico) a la concentración de la lejía límite, en contracorriente o en paralelo. El objeto de esta operación es la separación del exceso de soda cáustica, en caso dado de la glicerina, y de las impurezas solubles o dispersadas. El número de lavados y su intensidad dependen del grado de pureza que deberá tener el producto terminado.

it lavaggio
Questa operazione effettuata con soluzioni acquose di elettroliti, porta alla totale separazione della glicerina e delle impurezze dal sapone lavato dopo la salatura.

ne wassen
Operatie middels waterige electrolytoplossingen met het doel de afscheiding van de glycerine en de onzuiverheden uit de kernzeep te vervolmaken.

pl przemywanie
Operacja wykonywana przy pomocy wodnych roztworów elektrolitów, której celem jest oddzielenie resztek gliceryny i zanieczyszczeń z wysołu ściętego otrzymanego przez wysolenie.

659 **washing fastness, washing resistance**
fr solidité au lavage
de Waschbeständigkeit, Waschechtheit
el resistencia al lavado
it solidità al lavaggio
ne wasbestendigheid
pl trwałość na pranie, odporność na pranie

660 **washing fastness of colours**
fr solidité des couleurs au lavage, résistance de lavage des couleurs
de Waschechtheit der Farben
el resistencia de los colores al lavado, solidez de los colores al lavado
it solidità dei colori al lavaggio
ne wasbestendigheid der kleuren
pl odporność wybarwień na pranie, odporność kolorów na pranie

661 **washing instructions**
fr instructions pour le lavage
de Waschvorschrift
el instrucciones para el lavado, instrucciones para lavar
it istruzioni per il lavaggio
ne wasvoorschrift
pl przepis prania

662 **washing powder**
fr poudre à laver
de pulverförmiges Waschmittel
el polvo para lavar
it polvere per lavare
ne waspoeder
pl proszek do prania

663 **washing power**
Degree of aptitude of a surface active agent or of a detergent to promote detergency.

fr pouvoir détergent
Degré d'aptitude d'un agent de surface ou d'un détergent à promouvoir la détergence.

de Waschvermögen
Grad der Fähigkeit einer grenzflächenaktiven Verbindung oder eines Waschmittels das Phänomen der Waschwirkung hervorzurufen.

el poder detergente
Capacidad de un agente de superficie o de un detergente para promover la detergencia.

it potere detergente
Grado dell'attitudine di un tensioattivo o di un detergente a promuovere il fenomeno della detergenza.

ne waskracht
Vermogen van een oppervlakaktieve stof of van een wasmiddel om de waswerking te bevorderen.

pl zdolność piorąca
Zdolność związku powierzchniowo czynnego lub środka piorącego do działania piorącego.

664 **washing time**
fr durée de lavage
de Waschzeit, Waschdauer
el duración del lavado
it durata di lavaggio
ne wastijd
pl czas prania

665 **water and oil retaining capacity**
fr pouvoir de rétention pour l'eau et l'huile
de Wasser- und Öl-Rückhaltevermögen
el poder de retención de agua y aceite
it capacità di trattenere acqua ed olio
ne vermogen om water en olie vast te houden
pl zdolność zatrzymywania wody i oleju

666 **water fluoridation**
fr fluorisation de l'eau
de Fluorierung des Wassers
el fluorización del agua
it fluorizzazione dell'acqua
ne fluoridering van water
pl fluorowanie wody

667 **water hardness**
fr dureté de l'eau
de Wasserhärte
el dureza del agua
it durezza dell'acqua

ne hardheid van het water
pl twardość wody

668 wave set
fr mise en plis
de Einlegen der Welle
el marcado
it messa in piega
ne haargolfmiddel
pl układanie włosów

669 wave set powder
fr poudre pour mise en plis
de puderförmiges Einlegemittel
el polvos para el marcado
it polvere per la messa in piega
ne poedervormig haargolfmiddel
pl proszek do układania włosów

670 waving agent
fr produit pour permanente
de Dauerverformungsmittel, Dauerwellmittel, keratoplastische Verbindung
el producto para la permanente, agente ondulador
it prodotto per permanente
ne onduleermiddel
pl środek do ondulacji włosów, związek keratoplastyczny

671 waving procedure
fr procédé d'ondulation
de Wellvorgang
el procedimiento de ondulación
it processo di ondulazione
ne onduleermethode
pl proces ondulowania

672 way of washing, washing procedure
fr procédé de lavage, méthode de lavage
de Waschvorgang
el clase del lavado, manera del lavado, método del lavado, procedimiento del lavado
it processo di lavaggio, metodo di lavaggio
ne wasmethode
pl metoda prania, sposób prania, proces prania

673 wettability[1]
The ability of a surface to become wetted.
fr mouillabilité
Aptitude d'une surface à la mouillance.
de Benetzbarkeit
Fähigkeit einer Oberfläche, benetzt zu werden.

el humectabilidad
Capacidad de una superficie para ser mojada.
it bagnabilità
Attitudine di una superficie ad essere bagnata.
ne bevochtigbaarheid
Het vermogen van een oppervlak zich te laten bevochtigen.
pl zwilżalność
Zdolność powierzchni do zwilżania się.
[1] This definition corresponds to the French version, but in English usage definition No. 681 (wetting tendency) is the fundamental concept and should be given first consideration, the definitions Nos. 674, 673, 680 and 675 stemming from it, instead of vice versa.

674 wetting
In the special case of a surface active agent in solution: action corresponding to bringing into effect the properties of wetting tendency and wettability.
fr mouillage
Dans le cas particulier d'un agent de surface en solution: action correspondant à la mise en œuvre des propriétés de mouillance et de mouillabilité.
de Netzen
Für den speziellen Fall gelöster grenzflächenaktiver Stoffe: Vorgang, der durch das Netzvermögen der Lösung und die Benetzbarkeit der Oberfläche bewirkt wird.
el humectación, mojado
En el caso particular de un agente de superficie en disolución: efecto de la acción humectante de una disolución y de la humectabilidad de una superficie.
it bagnatura
Nel caso particolare delle soluzioni di tensioattivi: azione che fa si che si esplichino le proprietà della bagnabilità e dell'attitudine a bagnare.
ne bevochtigen
In het bijzondere geval van een oppervlakaktieve stof in oplossing: de werking die het resultaat is van bevochtigend vermogen en de bevochtigbaarheid.
pl zwilżanie
W szczególnym przypadku rozpuszczonego związku powierzchniowo czynnego proces będący wynikiem zdolności zwilżania roztworu i zwilżalności powierzchni.

675 wetting agent[1]
A substance which, when introduced

into a liquid, increases its wetting tendency.

fr agent mouillant, mouillant
Produit qui, introduit dans un liquide, augmente son aptitude à la mouillance.

de Netzmittel
Produkt, welches, wenn es einer Flüssigkeit zugesetzt wird, das Netzvermögen derselben vergrössert.

el producto humectante, humectante
Producto que introducido en un líquido, aumenta su capacidad para mojar una superficie.

it agente bagnante
Sostanza che sciolta o dispersa in un liquido ne aumenta l'attitudine a bagnare.

ne bevochtiger
Een stof die bij toevoeging aan een vloeistof het bevochtigend vermogen daarvan vergroot.

pl środek zwilżający
Substancja, która po wprowadzeniu do cieczy podwyższa jej zdolność do zwilżania.

[1] This definition corresponds to the French version, but in English usage definition No. 681 (wetting tendency) is the fundamental concept and should be given first consideration, the definitions Nos. 674, 673, 680 and 675 stemming from it, instead of vice versa.

676 wetting agent and dyeing oil
Product increasing the wetting power of the dye bath for the textile fabric. The dyeing oil generally confers on the articles dyed an additional brightening effect.

Note: These products are surface active agents or preparations comprising them, such as: alkylsulphates, alkylsulphonates, alkylarylsulphonates, fatty acid condensates, sulphated oils.

fr mouillant et huile pour teinture
Produit augmentant le pouvoir mouillant du bain tinctorial pour la matière textile. L'huile pour teinture donne en général à l'article teint un effet d'avivage supplémentaire.

Nota: Il s'agit d'agents de surface ou de préparations en comportant, tels que: alkylsulfates, alkylsulfonates, alkylarylsulfonates, condensats d'acides gras, huiles sulfatées.

de Färbereinetzmittel und Färbeöl
Produkt, das das Netzvermögen gegenüber den zu färbenden Textilien erhöht. Färbeöl verleiht dem Färbegut in der Regel noch einen zusätzlichen Avivageeffekt.

Anmerkung: Es handelt sich um grenzflächenaktive Stoffe oder Zubereitungen hieraus, wie Alkylsulfate, Alkylsulfonate, Alkylarylsulfonate, Fettsäurekondensationsprodukte und sulfierte Öle.

el humectante y aceite para la tintura
Producto que aumenta el poder humectante del baño de tintura. El aceite para tintura da en general al artículo teñido un efecto adicional de avivado.

Observación: Se trata de agentes de superficie, o de preparaciones que los contienen. Se utilizan alquilsulfatos, alquilsulfonatos, alquilarilsulfonatos, productos de condensación de ácidos grasos, aceites sulfatados.

it —

ne bevochtigingsmiddel
Produkten die het bevochtigend vermogen van het verfbad voor het weefsel vergroten.

pl środek zwilżający dla farbiarstwa
Produkt zwiększający zdolność zwilżania tkanin przez kąpiele farbiarskie. Niektóre z tych produktów dodatkowo nadają barwionym wyrobom efekt ożywienia.

Uwaga: Produktami takimi są związki powierzchniowo czynne takie jak alkilosiarczany, alkilosulfoniany, alkiloarylosulfoniany, produkty kondensacji kwasów tłuszczowych, siarczanowane oleje oraz preparaty zawierające te produkty.

677 wetting agent for the textile industry
Product which, when added to a solution, increases the wetting power of the latter. In the textile industry, it promotes the wetting and penetration of textiles by water or aqueous solutions.

Note: These products are surface active agents or preparations comprising them, such as: sulphated oils, fatty acid esters and amides, and also alkylsulphates, alcane sulphonates, fatty acid condensates, esters of alkylsulphosuccinic acid, oxalkylated products.

fr agent mouillant pour l'industrie textile
Produit qui, mis en solution, augmente le pouvoir mouillant de cette dernière. Dans l'industrie textile, il favorise le mouillage et la pénétration des textiles par l'eau ou les solutions aqueuses.

Nota: Il s'agit d'agents de surface ou de préparations en comportant, tels que: huiles sulfatées, esters et amides d'acides gras, ainsi que alkylsulfates, alkylsulfonates, condensats d'acides gras, esters sulfosucciniques, produits oxalkylés.

de Netzmittel für die Textilindustrie
Produkt, das — in Lösung gebracht — deren Netzkraft vergrössert. In der Textilindustrie begünstigt es das Benetzen und Durchdringen von Textilien mit Wasser oder wässrigen Lösungen.

Anmerkung: Es handelt sich um grenzflächenaktive Stoffe oder Zubereitungen aus diesen, wie sulfierte Öle, Fettsäureester oder -amide, ferner Alkylsulfate, Alkansulfonate, Fettsäurekondensationsprodukte, Alkyl-Sulfobernsteinsäureester, Oxalkylierungsprodukte.

el agente humectante para la industria textil
Producto que añadido a una disolución, aumenta su poder humectante. Favorece la humectación de los artículos textiles y la penetración en ellos del agua o de las disoluciones acuosas.
Observación: Se trata de agentes de superficie o de preparaciones que los contienen, tales como aceites sulfatados, ésteres y aminas de ácidos grasos, así como alquilsulfatos, alquilsulfonatos, productos de condensación de ácidos grasos, ésteres del ácido sulfosuccínico, ésteres y éteres de poliglicoles.

it —

ne bevochtiger voor de textielindustrie
Een produkt dat, toegevoegd aan een oplossing het bevochtigend vermogen hiervan vermeerdert. In de textielindustrie bevordert het de bevochtiging en het doordringen van water of waterige oplossingen in stoffen.
Opmerking: Deze produkten zijn oppervlakaktieve stoffen of preparaten die deze bevatten, zoals gesulfateerde oliën, vetzure esters en amiden, alsmede alkylsulfaten, alkylsulfonaten, vetzuurcondensaten, esters van alkylsulfobarnsteenzuur, polyglycolesters en -ethers.

pl środek zwilżający dla przemysłu włókienniczego
Produkt, który dodany do roztworu podwyższa jego zdolność zwilżania. W przemyśle włókienniczym podwyższa on zwilżanie i wnikanie wody lub roztworów wodnych do wyrobów włókienniczych.
Uwaga: Produktami takimi są związki powierzchniowo czynne takie jak siarczanowane oleje, estry i amidy kwasów tłuszczowych, alkilosiarczany, alkanosulfoniany, produkty kondensacji kwasów tłuszczowych, estry kwasu alkilosulfobursztynowego, produkty oksyalkilenowane oraz preparaty zawierające te związki.

678 wetting effect
fr mouillance
de Netzwirkung
el efecto humectante
it effetto di imbibizione
ne bevochtigende werking
pl działanie zwilżające, efekt zwilżania

679 wetting hysteresis
Under certain conditions, the work required to effect the introduction at slow and constant speed of a part of a solid surface into a liquid phase differs from that corresponding to the comparable withdrawal of the same part of the surface from the liquid phase. The difference between these two quantities, related to unit surface area, represents a hysteresis termed wetting hysteresis. It is expressed in joules $(J)^{1)}$ and is numerically equal to the difference between the advancing and receding wetting tensions.

fr hystérésis de mouillage
Dans certaines conditions, le travail nécessaire à la pénétration lente et à vitesse constante d'un élément d'une surface solide dans une phase liquide est différent de celui qui correspond à l'extraction, réversiblement comparable, du même élément de surface hors de la phase liquide. La différence entre ces deux quantités, rapportée à l'unité de surface, est représentative d'une hystérésis appelée hystérésis de mouillage. Cette grandeur s'exprime en joules $(J)^{1)}$ et est numériquement égale à la différence entre les tensions de mouillage rentrante et sortante.

de Benetzungshysterese
Die Arbeit, die nötig ist, um ein Flächenelement einer festen Oberfläche langsam und mit konstanter Geschwindigkeit in eine flüssige Phase einzutauchen, ist verschieden von der vergleichbaren Arbeit desselben Flächenelementes beim Herausziehen aus der flüssigen Phase. Die Differenz dieser Grössen, bezogen auf die Flächeneinheit, entspricht einer Hysterese, die Benetzungshysterese genannt wird. Sie ist die Differenz der fortschreitenden und rückläufigen Benetzungsspannung numerisch gleich und wird in Joule (J) ausgedrückt$^{1)}$.

el histéresis de mojado
En determinadas condiciones, el trabajo necesario para la penetración lenta y a velocidad constante de una superficie sólida en una fase líquida es distinto del que corresponde a su extracción, en condiciones reversibles comparables, de la misma fase líquida. La diferencia entre estos dos trabajos, representa una histéresis, llamada histéresis de mojado. Numéricamente es igual a la diferencia entre las tensiones de mojado entrante y saliente.

it isteresi di bagnatura
Il lavoro necessario per introdurre lentamente e ad una velocità costante una parte di superficie solida in una fase liquida è, in certe condizioni, diverso

da quello corrispondente all'estrazione nelle stesse condizioni, della stessa parte di superficie dalla stessa fase liquida. La differenza fra questi due lavori, riferita all'unità di superficie, rappresenta l'isteresi di bagnatura; si esprime in erg ed è numericamente uguale alla differenza fra la tensione di bagnatura progressiva e la tensione di bagnatura regressiva.

ne bevochtigingshysterese
Onder bepaalde omstandigheden is de spanning die vereist is om een deel van een vast fysisch oppervlak met een geringe en konstante snelheid in een vloeibare fase te brengen niet gelijk aan de vergelijkbare spanning waarmee ditzelfde deel van het oppervlak uit de vloeibare fase wordt getrokken. Het verschil tussen deze twee grootheden per oppervlak is een hysterese, genaamd "bevochtigingshysterese". Zij wordt uitgedrukt in J[1]) en is numeriek gelijk aan het verschil tussen de voortgaande en teruggaande bevochtigingsspanning.

pl histereza zwilżania
W pewnych warunkach praca wymagana dla powolnego wprowadzenia ze stałą prędkością części powierzchni ciała stałego do fazy ciekłej różni się od pracy potrzebnej do wyjęcia tego ciała z tej samej fazy ciekłej. Różnica pomiędzy tymi wielkościami w przeliczeniu na jednostkę powierzchni stanowi histerezę, zwaną histerezą zwilżania. Histereza ta wyrażana jest w dżulach (J) i liczbowo jest równa różnicy pomiędzy wstępującym i zstępującym napięciem zwilżania.
1) $1 \text{ J} = 10^7 \text{ erg}$.

680 wetting power[1])
The ability to wet.
fr pouvoir mouillant
Degré d'aptitude à la mouillance.
de Netzvermögen
Grad der Fähigkeit zur Benetzung.
el poder humectante
Capacidad para desarrollar la acción humectante.
it potere bagnante
Grado di attitudine a bagnare.
ne bevochtigend vermogen
Vermogen om te bevochtigen.
pl zdolność zwilżania
1) This definition corresponds to the French version, but in English usage definition No. 681 (wetting tendency) is the fundamental concept and should be given first consideration, the

definitions Nos. 674, 673, 680 and 675 stemming from it, instead of vice versa.

681 wetting tendency[1])
The tendency of a liquid to spread over a surface. A decrease in the contact angle between the solution and the surface is shown by an increase in wetting. A zero contact angle corresponds to spontaneous spreading.
fr mouillance
Tendance que possède un liquide à s'étaler sur une surface. Une diminution de l'angle de raccordement entre la solution et la surface se traduit par une augmentation de la mouillance. A un angle de raccordement nul correspond l'étalement.
de Benetzung
Bestreben einer Flüssigkeit, auf einer Oberfläche zu spreiten. Eine Verkleinerung des Randwinkels zwischen der Lösung und der Oberfläche äussert sich in einer Steigerung der Benetzung. Der Randwinkel Null entspricht der Spreitung.
el acción humectante
Tendencia de un líquido a extenderse sobre una superficie. Una disminución del ángulo de contacto entre la disolución y la superficie mojada se traduce por un aumento de la acción humectante. A un ángulo de contacto nulo, corresponde el esparcimiento.
it attitudine a bagnare
Attitudine di un liquido a spandersi su una superficie. La diminuzione dell'angolo di contatto fra la soluzione e la superficie considerata corrisponde ad un aumento di detta attitudine. Ad angolo di contatto nullo corrisponde lo spandimento.
ne bevochtiging
De neiging van een vloeistof zich over een oppervlak te verspreiden. Een verlaging van de kontakthoek tussen de vloeistof en het oppervlak resulteert in een vergroting van de bevochtiging. Een kontakthoek van 0° komt overeen met spontane spreiding.
pl zdolność zwilżania
Zdolność cieczy do rozprzestrzeniania się na powierzchni. Zmniejszenie kąta zwilżania pomiędzy roztworem a powierzchnią odzwierciedla się w zwięk-

szeniu zwilżania. Zerowy kąt zwilżania odpowiada samorzutnemu rozprzestrzenianiu (rozlewaniu) się cieczy.

[1] This definition corresponds to the French version, but in English usage definition No. 681 (wetting tendency) is the fundamental concept and should be given first consideration, the definitions Nos. 674, 673, 680 and 675 stemming from it, instead of vice versa.

682 wetting tension (symbol j)†
The force per unit of length resulting from the free wetting energy. It is expressed in newtons per metre (N/m)[1].

fr tension de mouillage (symbole j)†
Force par unité de longueur, résultante de l'énergie libre de mouillage. Elle s'exprime en newtons par mètre (N/m)[1].

de Benetzungsspannung (Symbol j)
Kraft pro Längeneinheit, die aus dem Vorhandensein der freien Benetzungsenergie resultiert. Sie wird in Newton pro Meter (N/m) ausgedrückt[1].

el tensión de mojado (líquido–sólido)
Fuerza por unidad de longitud, resultante de la energía libre de mojado. Numéricamente, es igual a la energía libre de mojado por unidad de superficie.

it tensione di bagnatura (liquido–solido)
Forza per unità di lunghezza, risultante dall'energia libera di bagnatura. E' numericamente uguale all'energia libera di bagnatura per unità di superficie e si esprime in dine per centimetro.

ne spanning van het bevochtigen (vloeistof–vaste stof)
De kracht per lengte-eenheid, voortkomende uit de vrije energie van het bevochtigen. Deze wordt uitgedrukt in N/m[1].

pl napięcie zwilżania
Siła na jednostkę długości wynikająca ze swobodnej energii zwilżania, wyrażona w niutonach na metr (N/m).

† See appedix
  Voir appendice
[1] 1 $N/m = 10^3$ dyn/cm.

683 whiteness degree
fr degré de blancheur, degré de blanc
de Weissgrad
el grado de blancura
it grado di sbianca
ne witheidsgraad
pl stopień białości, stopień bieli

684 winding oil
Product intended to make yarns suitable for winding and for subsequent textile operations, such as knitting, by making the yarns more flexible and slippery.

Note: These products are oils, or oils which can be emulsifiable in water, which can be prepared with the aid of surface active agents such as: oil-soluble oxalkylated products.

fr huile de bobinage
Produit destiné à rendre des fils aptes au bobinage et aux opérations textiles ultérieures, telles que le tricotage, en augmentant la souplesse et le glissant des fils.

Nota: Il s'agit d'huiles, ou d'huiles émulsionnables dans l'eau, pouvant être préparées à l'aide d'agents de surface tels que produits d'oxyalkylation oléosolubles.

de Spulöl
Produkt, das dazu bestimmt its, Garne spulfähig und durch Erhöhung der Geschmeidigkeit und Gleitfähigkeit für die weiteren textilen Arbeitsgänge, wie Wirkprozesse, geeignet zu machen.

Anmerkung: Es handelt sich um Öle oder wasseremulgierbare Öle, die unter Verwendung von grenzflächenaktiven Stoffen, z.B. öllöslichen Oxalkylierungsprodukten, zubereitet sein können.

el aceite de bobinado
Producto destinado para hacer los hilos aptos al bobinado y a las operaciones textiles posteriores, tales como el tricotaje, y para aumentar su suavidad y capacidad de deslizamiento.

Observación: Se trata de aceites o de aceites emulsionables en agua, que pueden ser preparados con agentes de superficie, tales como: ésteres o éteres de poliglicoles oleosolubles.

it —
ne spoelolie
Een eindavivage die de garens geschikt maakt voor het spoelen zodat een regelmatige spoelopbouw wordt verkregen.

pl olej stosowany przy przewijaniu przędzy
Produkt przystosowujący przędzę do przewijania oraz dalszych operacji włókienniczych takich jak np. dzianie. Efekt taki uzyskiwany jest poprzez nadanie przędzy większej elastyczności i śliskości.

Uwaga: Produktami takimi są oleje lub oleje emulgujące się w wodzie, otrzymane przy użyciu związków powierzchniowo czynnych takich jak rozpuszczalne w olejach produkty oksyalkilenowane.

Y

685 yellowing
fr jaunissement
de Vergilbung
el amarilleamiento
it ingiallimento
ne vergelen
pl żółknięcie

686 yellowish tinge
fr nuance jaunâtre
de Gelbstich
el matiz amarillento
it nuanza giallastra
ne gele verkleuring
pl zażółcenie, żółtawy odcień

Y

APPENDIX

Major divergencies of the ISO definitions as published in ISO R 862, in standard DIN 53 900 and by the IUPAC[1] from the C.I.T. definitions.

[1] "Pure and Applied Chemistry" 31 (1972) p. 611–613.

36 ISO 103
en antistatic agent
Product which, when applied to a textile article during or after processing, makes it possible to eliminate the disadvantages due to phenomena of static electricity.

Note: These products are generally surface active agents, for example: alkylsulphonates, alkylphosphates, alkylamines and their derivatives, and also the ethoxylation products of fatty acids, fatty alcohols, fatty amines, fatty amides, alkylphenols, and quaternary ammonium salts.

fr agent antiélectrostatique*
Produit qui, appliqué à un article textile en cours d'élaboration ou terminé, permet d'éviter les inconvénients dus aux phénomènes d'électrisation.

Note: Il s'agit en général d'agents de surface comme, par exemple: alkylsulfonates, alkylphosphates, alkylamines et leurs dérivés, ainsi que des produits d'éthoxylation d'acides gras, d'alcools gras, d'amines grasses, d'amides gras, d'alkylphénols et des sels d'ammonium quaternaire.
* Souvent, improprement appelé "Agent antistatique".

149 IUPAC
en curd soap
A soap curd is not a mesomorphic phase*[), but a gellike mixture of fibrous soap-crystals ("curd-fibres") and their saturated solution.

*) See appendix No. 395 Middle soap.

179 IUPAC
en detergent
A detergent is a surfactant (or a mixture containing one or more surfactants) having cleaning properties in dilute solution (soaps are surfactants and detergents).

192 ISO 120
en discharging agent
Product which, when added to a printing paste of discharge, makes it possible to discharge colour satisfactorily in the case of a dye which is difficult to discharge.

Note: These products are mainly based on derivatives of quaternary ammonium and ethoxylated amines.

fr adjuvant de rongeage
Produit qui, ajouté à une pâte d'impression de rongeage, permet d'obtenir un enlevage satisfaisant dans le cas d'une teinture difficile à ronger.

Note: Il s'agit essentiellement de produits à base de dérivés d'ammonium quaternaire et d'amines éthoxylées.

221 IUPAC
en emulsifying agent (emulsifier)
An emulsifier is a surfactant which when present in small amounts facilitates the formation of an emulsion, or enhances its colloidal stability by decreasing either or both of the rates of aggregation and coalescence.

277 IUPAC
en foaming agent
A foaming agent is a surfactant which when present in small amounts facilitates the formation of a foam, or enhances its colloidal stability by inhibiting the coalescence of bubbles.

282 ISO 22
en free energy of adhesion
The work required to achieve, in an isothermal, isobaric and reversible manner, a separation at the interface between two phases (liquid/solid) with the formation of a new free liquid surface of the same dimensions as the initial interface, is reflected in an increase in the free energy of the system. This energy is called *free energy of adhesion*. It is the sum of the free energy of wetting and the free surface energy. It is expressed in ergs*.

fr énergie libre d'adhésion
Le travail à fournir pour provoquer d'une façon isotherme, isobare et réversible une séparation à l'interface limitant deux phases (liquide/solide) avec formation d'une nouvelle surface liquide libre conservant les mêmes dimensions que l'interface initiale, se traduit par un apport d'énergie libre au système. Cette énergie est appelée *énergie libre d'adhésion*. Elle est la somme de l'énergie libre de mouillage et de l'énergie superficielle libre; elle s'exprime en ergs*.

DIN 2.2.2.3.
de freie Adhäsionsenergie
Die für die isotherm und reversibel durchgeführte Trennung zweier Phasen (flüssig–fest) an ihrer Grenzfläche anzuwendende Arbeit bewirkt eine Zunahme der freien Energie des Systems. Diese Energie wird freie Adhäsionsenergie genannt. Sie ist gleich der Summe der freien Benetzungsenergie und der freien Oberflächenenergie.

* The erg (symbol: erg) is the unit of the CGS system. The SI unit is the joule (J): 1 J = 10^7 erg.
* L'erg (symbole: erg) est l'unité du système CGS. L'unité SI est le joule (J): 1 J = 10^7 ergs.

286 ISO 23
en free energy of wetting
The work obtained when a surface is wetted in an isothermal, isobaric and reversible manner without changing the size of the free liquid surface, is reflected in a diminution in the free energy of the system. This part of the free energy is called *free energy of wetting*. It is expressed in ergs*.
fr énergie libre de mouillage
Le travail obtenu lorsqu'on mouille une surface d'une façon isotherme, isobare et réversible, sans modification simultanée de la grandeur de la surface liquide libre, se traduit par une diminution de l'énergie libre du système. Cette part d'énergie libre s'appelle *énergie libre de mouillage*; elle s'exprime en ergs*.
DIN 2.2.2.3.
de differentielle freie Benetzungsenergie (flüssig–fest)
Die Arbeit, die von System geleistet wird, wenn eine feste Oberfläche isotherm und reversibel bei konstantem Volumen ohne gleichzeitige Änderung der freien Flüssigkeitsoberfläche benetzt wird, kommt in einer Abnahme der freien Energie des Systems zum Ausdruck. Die negative Ableitung dieser freien Energie nach der benetzten Fläche wird differentielle freie Benetzungsenergie genannt.

* The erg (symbol: erg) is the unit of the CGS system. The SI unit is the joule (J): 1 J = 10^7 erg.
* L'erg (symbole: erg) est l'unité du système CGS. L'unité du système SI est le joule (J): 1 J = 10^7 ergs.

336 ISO 18
en interfacial tension
The force per unit of length, arising

from the free interfacial energy. It is numerically equal to the free interfacial energy per unit of interface and is expressed in dynes per centimetre*.
fr tension interfaciale
Force par unité de longueur, résultante de l'énergie interfaciale libre. Elle est numériquement égale à l'énergie interfaciale libre par unité d'interface, et s'exprime en dynes par centimètre*.

* The dyne per centimetre (dyn/cm) is the unit of the CGS system. The SI unit is the newton per metre (N/m): 1 N/m = 10^3 dyn/cm.
* La dyne par centimètre (dyn/cm) est l'unité du système CGS. L'unité SI est le newton par mètre (N/m): 1 N/m = 10^3 dyn/cm.

345 IUPAC
en Krafft point
Krafft point, symbol t_K (Celsius or other customary temperature), T_K, (thermodynamic temperature) is the temperature (more precisely, narrow temperature range) above which the solubility of a surfactant rises sharply. At this temperature the solubility of the surfactant becomes equal to the c.m.c. It is best determined by locating the abrupt change in slope of a graph of the logarithm of the solubility against t or $1/T$.

395 IUPAC
en middle soap
Concentrated systems of surfactants often form liquid crystalline phases, or mesomorphic phases. Mesomorphic phases are states of matter in which anisometric molecules (or particles) are regularly arranged in one (nematic state) or two (smectic state) directions, but randomly arranged in the remaining direction(s). Examples of mesomorphic phases are: neat soap, a lamellar structure containing much (e.g. 75%) soap and little (e.g. 25%) water; middle soap, containing a hexagonal array of cylinders, less concentrated (e.g. 50%), but also less fluid than neat soap.

505 ISO 42
en saponification
A chemical reaction permitting the separation of an ester into its constituent parts, acid and alcohol or possibly phenol, by the action of a base, with the formation of a salt from the acid. Saponification of fats produces soap.

fr saponification
Réaction chimique permettant de séparer les éléments constitutifs acide et alcool, ou éventuellement phénol, d'un ester, par l'action d'une base avec formation d'un sel aux dépens de l'acide. La saponification des corps gras conduit au savon. DIN 3.1.

de Verseifung
Chemische Reaktion, welche die Zerlegung eines Esters in seine Bestandteile, Säure, Alkohol (oder Phenol) durch Einwirkung einer Base ermöglicht und bei welcher ein Salz der Säure entsteht. Die Verseifung von Fetten führt zur Seife.

552 IUPAC
en solubilization
In a system formed by a solvent, an association colloid and at least one other component (the solubilizate), the incorporation of this other component into or on the micelles is called micellar solubilization, or, briefly solubilization. If this other component is sparingly soluble in the solvent alone, solubilization can lead to a marked increase in its solubility due to the presence of the association colloid. More generally, the term solubilization has been applied to any case in which the activity of one solute is materially decreased by the presence of another solute.

602 IUPAC
en surface active agent
A surface active agent (= surfactant) is a substance which lowers the surface tension of the medium in which it is dissolved, and/or the interfacial tension with other phases, and, accordingly, is positively adsorbed at the liquid vapour and/or at other interfaces. The term surfactant is also applied correctly to sparingly soluble substances, which lower the surface tension of a liquid by spreading spontaneously over its surface.

682 ISO 27
en wetting tension
The force per unit of length resulting from the free energy of wetting. It is numerically equal to the free energy of wetting per unit of surface. It is expressed in dynes per centimetre*.

fr tension de mouillage
Force par unité de longueur, résultante de l'énergie libre de mouillage. Elle est numériquement égale à l'énergie libre de mouillage par unité de surface, et s'exprime en dynes par centimètre*.

* The dyne per centimetre (dyn/cm) is the unit of the CGS system. The SI unit is the newton per metre (N/m): $1 \text{ N/m} = 10^3 \text{ dyn/cm}$.
* La dyne par centimètre (dyn/cm) est l'unité du système CGS. L'unité SI est le newton par mètre (N/m): $1 \text{ N/m} = 10^3 \text{ dyn/cm}$.

INDEXES

ENGLISH

abrading toothpowder 1
abrasive 2
acid cream 3
acid layer of the skin 4
acid mantle 5
acid mantle of the skin 4
active matter 6
active oxygen 7
adhesion energy, free 282
adsorption layer of surface
    active agents 8
advancing wetting angle 9
advancing wetting tension 10
aerosol foam waving
    compound 11
aerosol toothpaste 12
affinity to fibres 13
after-shave lotion
    (preparation) 14
after-taste 15
after-treating agent for
    prints 16
all-purpose cream 17
all-purpose washing agent 18
alopecia 372
alpha phase 19
amide formation 20
amino acid 21
amino dye 22
ammoniated toothpaste 23
ammonium thioglycolate 24
amount of foam 25
amphiphilic product 26
ampholytic surface active
    agent 27
ampholytics 28
amphoteric surfactants 29
anaeshetic
    topical 641
analgesic 30
ancillary (for surface active
    agents) 31
anionic surface active agent 32
anionics 33
antibacterial action 56
antidandruff agent 34
antidandruff shampoo 35
antielectrostatic agent 36
anti-enzyme toothpaste 37
antifoaming agent
    (antifoamer) 38
antifoaming agent for the
    textile industry 39
anti-perspirant 40
anti-redepositing power 41
anti-redeposition agent 42
antiseptic hand cream 43
anti-wrinkle cream 44
apparent density 45

apparent volume 46
application technique 47
aqueous emulsion 48
artificial soil 49
artificially soiled test cloth 50
ash content 51
atomisation 52
atomiser
    centrifugal disk 99
    spinning disk 99
autoxidation 53
available oxygen 54

baby cream 55
bactericidal action 56
baldness 57
ballpoint applicator 58
bar
    soap 59
    synthetic 616
base coat 60
bath
    bleaching 76
    dyeing 213
    ratio 61
    soap 62
    washing 656
bath additive
    spinning 558
beads (hollow) 63
beauty lotion
    tonic 636
beauty mask 64
beta phase 65
binder 66
binding agent 66
biodegradability 67
biodegradable surface active
    agent 68
biodegradation 69
    degree 163
blanching of the hair 295
bleach, to 70
    chlorine 106
    cream 71
    effectiveness of the optical
      bleach 217
    nail 402
bleaching 72
bleaching agent 73
    chemical 105
    optical 429
bleaching assistant 74
bleaching bath 76
bleaching compound
    oxygenated 434
bleaching efficiency 75
bleaching liquor 76
bleaching process 77

bleaching product 73
block of colouring 577
boiler
    soap 535
boiling 78
booster 79
    foam 268
    suds 272
breakage
    foam 350
breath freshener 80
brightener
    optical 429
brightening agent 81, 429
brightening effect 82
brush
    lipstick 364
    shaving 517
    tooth 637
brushless shaving cream 83
bubble 84
buffer capacity of the skin 85
builder (for surface active
    agents) 86
bulk density 87

caking 88
calcification of teeth 89
calcium chelating power 90
calcium hardness 91
calcium sequestering power 90
capillary activity 92
carbonizing assistant 93
carboxymethylcellulose
    (CMC) 94
caries
    dental 167
carious spot 95
cationic surface active agent 96
cationics 97
centrifugal disk atomisation 98
centrifugal disk atomiser 99
centrifugal disk spray wheel 99
change in shade 100
chelate 101
chelating agent 102
chelating power 103
    calcium 90
chelation 104
chemical bleaching agent 105
chlorine bleach 106
chlorophyll toothpaste 107
clarification
    temperature 621
cleaner
    denture 169
cleaning 189
    effect 108
    general 288

FRANÇAIS

DEUTSCH

ESPAÑOL

ablandamiento 540
ablandar 539
abrasivo 2
abrillantamiento 376
acción
  antibacteriana 56
  bactericida 56
  detergente 180
  detergente sobre la vajilla 195
  deterioradora para la fibra 155
  humectante 681
  mecánica 387
aceite
  de bobinado 684
  de corte 153
  detergente 185
ácido
  ditioglicólico 207
  tioglicólico 630
  tioláctico 631
aclarar la piel 355
actividad
  de superficie 604
  interfacial 335
  superficial 604
adherencia de la suciedad 544
aditivo 31
  para el baño de hilatura 558
  para la disolución de hilatura 560
afección dermatológica 526
afinidad
  de la suciedad para la fibra 544
  para las fibras 13
agente
  anti-caspa 34
  anti-reductor 482
  antielectrostático 36
  antiespumante para la industria textil 39
  blanqueante óptico 429
  contra la caspa 34
  de anti-redeposición 42
  de avivado 81
  de blanqueo con oxígeno 434
  de blanqueo oxigenado 434
  de blanqueo químico 105
  de desmontado 581
  de disolución y/o de dispersión de colorantes 554
  de disolución y/o de dispersión de pigmentos 555
  de ensimaje 559
  de humidificación de los hilados 251
  de mejoración de espuma 272
  de preparación 467

agente
  de superficie 602
  de superficie anfólito 27
  de superficie aniónico 32
  de superficie bio-resistente 420
  de superficie biodegradable 68
  de superficie catiónico 96
  de superficie no iónico 423
  del aclarado 493
  detergente para la industria textil 184
  dispersante 201
  emulsionante para la industria textil 222
  engrasador 600
  engrasante 600
  espesante 66
  humectante para la industria textil 677
  igualador 353
  ondulador 670
  para el tratamiento posterior de estampados 16
  para et tratamiento posterior de tinturas 212
  peptizante 445
  productor de uniones transversales 147
  protector de fibras 253
  protector de la piel 530
  protector del sol 596
  pulimentador 459
  que aumenta la espuma 272
  quelatante 102
  quitamanchas 562
  secuestrante 511
  solubilizante 553
  suavizante 541
  tensioactivo 602
aglutinación 88
aglutinante 66
agua
  blanda 538
  de aclarar 494
  dentifrica 366
  dulce 538
  dura 311
  medianamente dura 397
  moderadamente dura 397
  oxigenada 324
  poco dura 538
ahuecado 613
alargado 580
almacenaje 578
almacenamiento 578
alopecia 372
altura de la espuma 271
amarilleamiento 685

amidificación 20
amino-ácido 21
analgésico 30
análisis granulométrico 438
anestésico local 641
ángulo
  de contacto 134
  de mojado entrante 9
  de mojado saliente 477
antiperspirante 40
aplicador de punta redonda 58
apreciación del "tacto" 339
aroma 262
ataque de la fibra 155
atezamiento 618
atomización 52
  en caliente 567
  por calor 567
  por disco rotativo 98
  por disco giratorio 98
  por enfriamiento 563
atomizador de disco rotativo 99
aumentador de espuma 349
autoxidación 53
auxiliar
  de batanado 287
  de blanqueo 74
  de carbonizado 93
  de corrosión 192
  de desencolado y de eliminación de espesantes de estampación 177
  de encolado 523
  de mercerizado 391
  para el apresto 260
  para el descrudado y la hidrofilización 342
bajos 418
baño
  de blanqueo 76, 373
  del lavado 656
  del sol 592
  para teñir 213
barniz base 60
barra de jabón de afeitar 520
barrita 576
  desodorante 172
base
  de polvos 142, 463
  para las uñas 60
bigudí 150
biodegradabilidad 67
biodegradación 69
blanqueador óptico 429
blanquear del cabello 295
blanqueo por cloro 106
brillo 374
brocha 517

ITALIANO

NEDERLANDS

POLSKI